Business Telecom Systems

A Guide to

Choosing

the Best

Technologies

and Services

BY KERSTIN DAY PETERSON

Business Telecom Systems

Published by CMP Books
An Imprint of CMP Media Inc.
12 West 21 Street
New York, NY 10010

ISBN 1-57820-041-5

For individual orders, and for information on special
discounts for quantity orders,
please contact:

CMP Books
6600 Silacci Way
Gilroy, CA 95020
Tel: 800-LIBRARY or 408-848-3854
Fax: 408-848-5784
Email: telecom@rushorder.com

Distributed to the book trade in the U.S. and Canada by
Publishers Group West
1700 Fourth St., Berkeley, CA 94710

Manufactured in the United States of America

Contents

Telecom Basics

Communications Systems and Hardware

An easy-to-use list of key systems PBXs and communications servers,
by manufacturer

Major phone system and comms server manufacturers'
contact information

Cabling and Wiring

Adjuncts and Add-Ons

Provisioning and Transport Services

Reference and Templates

Telecom Basics

Introduction

The telecommunications infrastructure that exists in a handful of high-income countries includes 71% of the world's phone lines and supports only 15% of the world's population. The least and lesser-developed countries of the world, with more than 77% of world population, have only 5% of the world's phone lines. More than half of the world's nearly six billion people have never even used a phone.

Telecommunications equipment giant Lucent Technologies Inc. predicts the global market for communications systems and services will grow 14.5% annually to $650 billion by 2001.

It's impossible to predict what the telecom market will look like in three to five years. The technological and regulatory changes underway are unprecedented. One thing you can count on is that the technology will continuously improve and that technology and services will spring up, and quickly.

Bringing phone service into a business is basically twofold. There's the provisioning, the outside part (how you connect to everyone else in the world you want to communicate with) and there's the equipment you use to make the connections.

There are exceptions to this, like Centrex service (for more see that chapter), but basically, you need a way to communicate inside your enterprise and from your enterprise to the outside world, equipment-wise, and a way to connect to the outside world, service-wise.

This book tries to make some sense of the industry to those not familiar to it, in a no-nonsense way so that it is easier for you to buy and equip your office telecommunications.

There is an additional section about a combination technology, Computer Telephony that impacts making and taking call in ways you should know about, even if your office is very small.

Computer telephony is an end-to-end digital technology that not only can let you make regular voice calls, it also lets you make and receive Internet, video, fax and e-mail calls, too. Computer telephony is also the technology behind phone-system related products like voice mail, auto attendant and ACD, or Automatic Call Distribution, a technology that's used when there are groups of people answering or making a lot of the same kind of calls.

The type and number of trunks or lines you use to bring in and take out your phone calls (or data) has something to do with the kind of equipment you need to handle the calls. More lines or digital lines mean bigger or fancier (or even just different), equipment and usually more money, on a per-station basis.

There are lots of ways you can answer and make basic voice phone calls. From technologically basic to complicated, they are:

POTS (Plain Old Telephone Service)

It's probably what you have at home; a one-line phone with a regular old number from the phone company. One-person offices can get away with a single line set and an answering machine or a voicemail box from the phone company. A fancier phone, with a hold button, perhaps, or more features from the phone company (see Centrex) would be top-of-the-line cheapest.

SOHO (Small Office, Home Office)

If you need more than one but not more than four regular outside lines, the next-cheapest way to get phone service is by using phones that can share lines. SOHO is a general term that refers to a one-person or very small business. If you work out of your home or have a tiny office that probably won't grow, you can get away very cheaply with SOHO, and still have a lot of the features that bigger systems offer.

Systems that are called SOHO can be KSUless phones or small key systems. The term doesn't usually differentiate between these types of systems, but the difference is pretty important in terms of how the equipment works and how it is wired together.

KSUless

KSUless systems are phones that are smart enough by themselves to form a system. There's no central computer (KSU, Key Service Unit) or processing unit. KSUless systems can give your tiny enterprise some really useful features for $100 to $150 per phone. You can buy them at office supply stores. And they're easy to install, because the phones can usually be plugged right into your existing jacks.

The biggest disadvantage to SOHO KSUless systems is their lack of expandability. Understand that the money spent on fancy phones will be forfeit when your company grows to over three or so people. KSUless systems can normally handle only up to four numbers.

Phone-only systems are also usually lacking in more sophisticated telephony tricks like voice mail, paging and music on hold.

Key Systems

There's a big difference between SOHO Ksuless systems and small key systems, both in the way they work and in the way they're installed. Key systems have a dedicated computer-controlled system unit, or KSU (Key System Unit). They are connected differently than KSUless systems, in that a wire has to be installed from the KSU to each phone, directly. Key systems sold for the SOHO market usually range from two lines to six lines and four phones to eight or twelve phones.

Private Branch Exchange (PBX)

The difference between a key system and a PBX (Private Branch Exchange) is basically one of size. Although, until recently, key systems were called key systems because they had significantly lesser capabilities than PBXs, small systems are now sporting abilities that used to be reserved for bigger companies and bigger bank accounts.

The smallest two or three or four or six line key system is often feature-short compared to PBXs designed for corporate use, but size is not a hard and fast rule. If you're expecting to grow and expand madly, or want fancy lines from the telephone company, or want to get fax, email and voice messages on your computer screen, you can buy a bigger-capacity system and only a few expansion cards for lines and phones and you'll have a small, feature rich (abeit more expensive) system.

In addition to centralized processing (done by the KSU), key systems and PBXs let you use different types of phone lines that might make running your business easier. They have voicemail that's included, or there are voicemail systems that you can buy that will integrate easily with the key or PBX system. (See the voicemail chapter for more about voicemail-phone system integration.)

Key systems and all PBXs distribute calls over the incoming and outgoing lines so that you can have fewer lines than people and can hook up phones with as few as one pair of wires. They have a brain in the form of a central process unit. And the bigger the brain, the bigger the system and the more features it'll have. PBXs are the biggest and smartest of phone systems.

Communications Servers

Comms servers switch phone calls like key systems and PBXs, using Computer Telephony (CT) technologies to integrate computer and telephone functions. They're usually a PC with voice cards running on the Windows NT operating system.

At this writing, comms servers can handle up to about 400 phones. Comms servers are limited in the number of ports they support, although the upper limit is being pushed very hard.

Provisioning

For the first time ever, you have a reliable alternative for local telephone service. What's more, you can now (or soon will be able too) obtain local, long distance, and Internet service from a single phone company. Best of all, both of these are or will be available at lower prices, bundled with better customer service.

Provisioning is the other side of the telecommunications function. Your CPE (Customer Premise Equipment) hands off any traffic (calls) leaving your building to the phone company, or local exchange carrier (LEC).

The LECs carry traffic within the local area and hand calls (with competition, these calls could be either local, local/long distance or long distance) off to an inter-exchange carrier (IXC). At the other end of a local/long distance or long distance call the IXC hands the call to the LEC to deliver it to the distant customer's premises.

The Local Loop is generally what connects your system to the nearest Central Office. It includes the copper cable, fiber optics, pole lines, conduits, terminals, and other facilities that deliver signals from the central office to your premises and vice versa.

Local switching equipment is the switching matrix at the Central Office (CO) of your LEC that connects lines and trunks together to establish a communications session. This is the phone company's phone system.

Local trunking is the network of circuits running between local central offices and between local and long distance switching systems or carriers.

Long distance systems are a combination of switching systems and circuits that haul telephone traffic between local exchange areas.

Provisioning is probably the most difficult part of buying and installing a phone system. The Telecommunications Act of 1996 de-regulated local service so that there are new companies offering local service and long distance carriers are local vendors now, too. There are also a lot of new products, most of them digital. Eventually, the competition will be better for consumers, but there will be a lot of confusion in the meantime.

The History of Telephones and Telecommunications

Telephone comes from the Greek word tele, meaning from afar, and phone, meaning voice. A telephone is a device that conveys sound over a distance.

On March 10, 1876, in Boston, Massachusetts, Alexander Graham Bell invented the telephone. Thomas Watson fashioned the device itself; a crude thing made of a wooden stand, a funnel, a cup of acid, and some copper wire.

The courts awarded Bell one of the most valuable patents in American history, a patent that made him rich and enabled one of America's largest corporations.

Bell succeeded because he understood acoustics, the study of sound, and something about electricity. Other inventors knew electricity well but little of acoustics.

The Beginnings of Telephony

A real telephone could not be invented until the electrical age began, and even then it didn't seem desirable. The electrical principles needed to build a telephone were known in 1831 but it wasn't until 1854 that Bourseul proposed transmitting speech electrically. And it wasn't until 22 years later in 1876 that the idea became a reality.

In 1729 the English chemist Stephen Gray transmitted electricity over a wire, sending charges almost 300 feet over brass wire and moistened thread. An electrostatic generator powered his experiments, one charge at a time.

In 1746 Dutchman Pieter van Musschenbroek and German Ewald Georg von Kleist independently developed the Leyden jar, a battery or condenser for storing static electricity. Named for the city of its invention in Holland, the jar was a glass bottle lined inside and out with tin or lead. The glass between the metal sheets stored a strong electrical charge that could kept for a few days and transported.

In 1800 Alessandro Volta developed the first battery. Volta's battery provided sustained low power electrical current. Chemically based, the battery improved quickly and became an electrical source for other experimenting.

Batteries got more reliable, but they still couldn't produce the power needed to work machinery, light cities, or provide heat. And although batteries would work telegraph and telephone systems, and still do, transmitting speech required understanding two related concepts, electricity and magnetism.

In 1821 Michael Faraday got a weak current to flow over a wire revolving around a permanent magnet. In other words, a magnetic field caused or induced an electric current to flow in a nearby wire. This was the world's first electric generator. Mechanical energy could now be converted to electrical energy.

The simple act of moving a conductor caused current to move and turned mechanical energy into electrical energy. Move or rotate a wire fast enough and things could begin to happen. Faraday worked through different electrical problems in the next ten years, eventually publishing his results on induction in 1831.

In 1830 Professor Joseph Henry transmitted the first practical electrical signal, shortly after he'd invented the first efficient electromagnet. He also made unpublished conclusions about induction similar to Faraday's. Henry proved that electromagnetism could be used to communicate.

In his classroom, Henry demonstrated the forerunner of the telegraph. He wrapped an iron bar with several feet of wire and built an electromagnet. A pivot mounted steel bar sat next to the magnet, and a bell stood next to the bar.

From the electromagnet Henry strung a mile of wire around the inside of the classroom, completing the circuit by connecting the ends of the wires at a battery. The steel bar swung toward the magnet striking the bell at the same time. Breaking the connection released the bar, freeing it to strike the bell again.

Joseph Henry did not pursue electrical signaling research himself, but he helped someone who did: Samuel Finley Breese Morse. Joseph Henry and Morse built a telegraph relay or repeater that allowed long distance operation.

Samuel Morse invented the first practical telegraph in 1746, applied for its patent in 1838 and was granted the patent in 1848. Morse was captivated by electrical experiments although not an inventor by profession.

In 1832 he heard of Faraday's recently published work on inductance, and was given an electromagnet at the same time to ponder over.

Morse's system used a key to make or break the electrical circuit, a battery to produce power, a single line joining one telegraph station to another and an electromagnetic receiver or sounder that upon being turned on and off, produced a clicking noise. He completed the package by devising the Morse code system of dots and dashes.

The telegraph helped unite the country and eventually the world. Telegraphy replaced the Pony Express, clipper ships, private messengers and all communications any slower than instantaneous. Service was limited to Western Union offices or large firms but convenience was hardly a problem when communicating over long distances was now instant.

The success of telegraph made inventors' thoughts turn to transmitting speech over a wire.

In 1854 the Belgian-born French inventor and engineer Charles Bourseul wrote about transmitting speech electrically using a flexible disk that made and broke an electrical connection, reproducing sound.

The Inventors: Gray and Bell

In the early 1870s Elisha Gray, Alexander Graham Bell, and many others were working on a multiplexing telegraph that could send several messages over one wire simultaneously. Such an instrument would greatly increase traffic without the telegraph company having to install more lines. As it turned out, the desire to invent the multiplexing telegraph turned into the invention of something very different.

Elisha Gray was born in 1835 in Barnesville, Ohio. His first telegraph-related patent came in 1868. An expert electrician, he co-founded Gray and Barton, makers of telegraph equipment. The Western Union Telegraph Company (funded by the Vanderbilts and J.P. Morgan), bought a one-third interest in Gray and Barton in 1872. Gray and Barton was renamed the Western Electric

Manufacturing Company. Transmitting speech was an interesting goal but not a lifetime passion for Gray.

Alexander Graham Bell, conversely, became consumed with inventing the telephone. Born in 1847 in Edinburgh, Scotland, Graham was raised in a family involved with music and the spoken word. His mother painted and played music. His father originated a system called visible speech that helped the deaf to speak. His grandfather was a lecturer and speech teacher. His entire education and upbringing revolved around the mechanics of speech and sound.

In 1870 Bell's father moved his family to Canada after loosing two sons to tuberculosis, hoping the Canadian climate would be safer. In 1873 Alexander Bell became a vocal physiology professor at Boston College and taught the deaf the visual speech system. In his spare time he worked on what was called a harmonic or musical telegraph.

Bell, familiar with acoustics, thought that he could send several telegraph messages at once by varying their musical pitch. The harmonic telegraph proved simple to visualize but almost impossible to build. Bell labored over the harmonic telegraph into the spring of 1874 when, at a friend's suggestion, he started working on a teaching aid for the deaf, a disgusting device called the phonoautograph. It was made out of a dead man's ear.

Speaking into the device caused the ear's membrane to vibrate and then move a lever. The lever wrote a wavy pattern representing the speech on smoked glass.

Bell fixated on how a thin membrane like the human ear's could make a much heavier lever work. He thought he could make a membrane work in telephony by using it to vary an electric current in intensity with speech. The current could then replicate the speech using another membrane. It took him another two years to figure out how to apply the principle.

Reaching for the Phone

Bell continued harmonic telegraph work through the fall of 1874. He wasn't showing a lot of progress but his tinkering got some attention. Gardiner Greene Hubbard, a prominent Boston lawyer and the president of the Clarke School for The Deaf, became interested in Bell's experiments. He and George Sanders, a wealthy Salem businessman, both gambled that Bell would get his harmonic telegraph to work. The three men signed a formal agreement in February, 1875 that gave Bell financial backing in return for equal shares from any patents Bell developed.

Bell's experimenting picked up quickly with the help of a talented machinist named Thomas A. Watson. Bell feverishly pursued the harmonic telegraph his backers wanted and at the same time, the telephone, now his real interest.

On March 1, 1875, Bell met with Joseph Henry, the great scientist and inventor, then Secretary of the Smithsonian Institution. (Henry, remember, pioneered electromagnetism and helped Morse with the telegraph.) Uninterested in Bell's telegraph work, Henry urged Bell to drop all other work and get on with developing the telephone. Bell took his advice.

On June 2, 1875, Bell and Watson were testing the harmonic telegraph when Bell heard a sound come through the receiver. Instead of transmitting a pulse, which he couldn't get it to do, anyway, the telegraph transmitted the sound of Watson plucking one of many differently-tuned springs.

Their telegraph, like all others, turned current on and off. But this time, a contact screw was set too tightly and allowed the current to run continuously, the essential element needed to transmit speech.

Bell figured out what happened and the next day had Watson build a telephone based on continuous current. The Gallows telephone, so called for its distinctive frame, substituted a diaphragm for the spring.

It didn't work. No speech could be transmitted, only a few sounds. Discouraged, tired, and running low on funds, Bell's experimenting slowed through 1875.

During the winter of 1875 and 1876 Bell continued experimenting while writing a telephone patent application. Although he hadn't developed a successful telephone, he felt he could describe how it could be done. With his ideas and methods protected he would focus on making a working model.

Fortunately for Bell, the Patent Office had dropped its requirement that a working model accompany a patent application in 1870. On February 14, 1876, Bell's attorney filed his patent application, only hours before Elisha Gray filed his Notice of Invention for a telephone.

Mystery still surrounds Bell's application and what happened that day. Most peculiar is that the key point to Bell's application, the principle of variable resistance, was scrawled in a margin, looking like an afterthought.

It's been contended, but never proven (despite some 600 lawsuits that would eventually challenge the patent), that Bell was told of Gray's Notice and allowed to change his application.

Telephonic Success

Finally, on March 10, 1876, one week after his patent was allowed, Bell succeeded in transmitting speech. Again, by coincidence or conspiracy, Bell used an idea he hadn't outlined in his patent or even tried before, the liquid transmitter. The liquid transmitter concept was outlined in Gray's Notice, however.

The Watson-built telephone looked odd and acted strangely. Bellowing into the funnel caused a small disk or diaphragm at the bottom to move. This disk was, in turn, attached to a wire floating in acid in a metal cup. Two wires attached to the cup connected it to the distant receiver. As the wire moved up and down in the acid it changed the resistance in the liquid. The varying current was then sent to the distant receiver.

The central claim of Bell's patent was that undulating current was the best method to transmit sound, as opposed to the on-off current commonly used in telegraphy. The undulatory current preserved the gradual changes in intensity produced by speech or musical tones. This is what is called analog signaling.

The Coming of Telephones to America

Early telephones were voice powered. No battery or line current charged the instrument. Later, a transmitter worked by itself, producing a weak current when spoken into. Good transmission only happened if the users were shouting and when distances measured in hundreds of yards. Bell and Watson managed a long distance call on October 9, 1876, a distance of only two miles.

Promoting and developing the telephone proved far harder than Hubbard, Sanders, or Bell expected. Despite Bell's patent, broadly covering the entire subject of transmitting speech electrically, many companies sprang up to sell telephones and telephone service.

Other people filed applications for telephones and transmitters after Bell's patent was issued. Most claimed Bell's patent couldn't produce a working telephone or that they had a prior claim. Litigation loomed. Fearing financial ruin, late in 1876 Hubbard and Sanders offered their telephone patent to Western Union for $100,000. Western Union refused.

On April 27, 1877 Thomas Edison filed a patent application for an improved transmitter, a device that made the telephone practical. A major accomplishment, Edison's patent claim was declared in interference to a Notice of Invention for a transmitter filed just two weeks before by Emile Berliner. This conflict was not resolved until 1886. Edison produced the transmitter while the matter was disputed, starting late in 1877.

Bell used an improved transmitter invented by Francis Blake. On July 9, 1877 Sanders, Hubbard, and Bell formed the first Bell telephone company. Each assigned their rights under four basic patents to Hubbard's trusteeship.

Against tough criticism, Hubbard decided to lease telephones and license franchises, instead of selling them. This had enormous consequences. Instead of making money quickly, dollars would flow in over months, years, and decades. It proved a wise enough decision to sustain the Bell System for over a hundred years.

The Earliest Local Competition

In September, 1877 Western Union changed its mind about telephony. They saw it would work and they wanted in, especially after a subsidiary of theirs, the Gold and Stock Telegraphy Company, ripped out their telegraphs and started using Bell telephones.

Rather than buying patent rights or licenses from the Bell, Western Union bought patents from others and started their own telephone company. They were not alone. At least 1,730 telephone companies organized and operated in the 17 years Bell was supposed to have a monopoly. These earliest local competitors either disagreed with Bell's right to the patent, ignored it altogether, or started their own phone company because Bell would not provide service to their area.

Western Union began entering agreements with Gray, Edison, and Amos E. Dolbear for their telephone inventions. In December, 1877 Western Union created the American Speaking Telephone Company. A tremendous selling point for their telephones was Edison's improved transmitter.

Bell Telephone was worried about competing with Western Union, since there were only 3,000 Bell phones installed by the end of 1877. Western Union, on the other hand, had 250,000 miles of telegraph wire strung over 100,000 miles of route. If not stopped, Western Union would have an enormous head start on making telephone service available across the country.

Western Union was realistically the world's largest telecom company then, with an unchallenged monopoly on telegraph service, however, Bell's shrewd Boston lawyers filed suit against them.

Switchboards are Vogue

On January 28 1878 the first commercial switchboard began operating in New Haven, Connecticut. It served 21 telephones on eight lines; many people were on a few party lines. On February 17 Western Union opened the first large city exchange in San Francisco. Phones were no longer wired point-to-point; subscribers could talk to others on different lines. The public switched telephone network was born.

Also in 1878 the Butterstamp telephone came into use. This telephone combined the receiver and transmitter into one hand-held unit. You talked into one end, turned the instrument around to listen to the other end. People got confused with this clumsy arrangement, and consequently, a telephone with a second transmitter and receiver unit was developed in the same year. You could use either one to talk or listen and you didn't have to turn the phone around. This wall set used a crank to signal the operator.

These early phones were still voice powered as was the original telephone, that is, no battery or external power helped speech get to the other party or a switchboard. On August 1, 1878 Thomas Watson filed for a ringer patent that was very like Henry's old classroom doorbell.

A hammer operated by an electromagnet struck two bells. Turning a crank on the calling telephone spun a magneto, producing a ringing current. Before this invention, subscribers used a crude thumper to signal the called party, hoping someone would be around to hear it. The ringer was immediately successful.

Subscribers grew steadily but slowly. Sanders had invested $110,000 by early 1878 without any return. He located a group of New Englanders willing to invest but unwilling to do business outside their area. Badly needing the funding, the Bell Telephone Company reorganized in June, 1878 (10,755 Bell phones were in service), and formed a new Bell Telephone Company, the forerunner of the strong regional Bell companies to come..

In early 1879 the company reorganized once again, under pressure from patent suits and from competition from other companies selling phones with Edison's superior transmitter. William H. Forbes was elected to head the board of directors and restructured it to embrace all Bell interests into a single company, the National Bell Company, which was incorporated on March 13, 1879.

On November 10, 1879 Bell won its patent infringement suit against Western Union in the United States Supreme Court. In the resulting settlement, Western Union gave up its telephone patents and the 56,000 phones it managed, in return for 20% of Bell rentals for the 17 year life of Bell's patents. It retained the telegraph business.

This decision allowed National Bell to enlarge. A new incarnation, the American Bell Company, was created on February 20,

1880, capitalized with over $7 million dollars. Bell now managed 133,000 telephones.

Chief Operating Officer Theodore Vail began building the Bell System, regional companies offering local service, plus a long distance company providing toll service and a manufacturing arm providing equipment.

The manufacturer was a previous company rival. In 1880 Vail started buying Western Electric stock and took controlling interest on November, 1881. The takeover was consummated on February 26, 1882, with Western Electric giving up its remaining patent rights as well as agreeing to produce products exclusively for American Bell.

On July 19, 1881 Bell was granted a patent for the metallic circuit, the concept of two wires connecting each telephone. Until that time a single iron wire connected telephone subscribers, just like a telegraph circuit. A conversation works over one wire since grounding each end provides a complete path for an electrical circuit.

Houses, factories and the telegraph system were all grounding their electrical circuits using the same earth the telephone company employed, introducing a huge amount of static. A metallic circuit, on the other hand, used two wires to complete the electrical circuit, avoiding the ground altogether and thus providing a better-sounding call.

It was not until 10 years later that Bell started converting grounded circuits to metallic ones. And it took another ten years to complete the project.

Long Distance is Born

In 1885 Vail formed his long distance telephone company, AT&T. Capitalized on only $100,000, American Telephone and

Telegraph provided long distance service for American Bell.

Only local telephone companies operating under Bell-granted licenses could connect to AT&T's long distance network. Vail thought this would continue the Bell System's virtual monopoly after its key patents expired in the 1890s, reasoning that the independents would not be able to compete since they would be isolated without long distance lines.

With only Bell-licensed companies providing local service, his Western Electric manufacturing equipment and AT&T long distance, Vail's structuring of the Bell System was now complete. His job done, in September 1887 Vail resigned from American Bell.

In 1889 the first public coin telephone came into use in Hartford, Connecticut. The first payphones were attended, with payment going to someone standing nearby.

In 1892 Bell controlled 240,000 telephones, but independents were coming on fast by using better technology. The first automatic dial system began operating that year in La Porte, Indiana. The central office switch worked in concert with a similar switch at the subscriber's home, operated by push buttons.

Central Office Switching

Patented in 1891 by Almon B. Strowger, the Step by Step or SXS system replaced the switchboard operator for placing local calls. People could dial the number themselves. This required different kinds of telephones and eventually models with dials. A.E. Keith, J. Erickson and C.J. Erickson later invented the rotating finger-wheel needed for a dial.

The first dial telephones began operating in Milwaukee's City Hall in 1896. Independents were quick to start using the new switch and phones. The Bell System however, did not embrace this switch or automation in general, indeed, a Bell franchise commonly removed step-by-step switches and dial telephones from territories it bought from independent telephone companies. Not until 1919 did the Bell System start using Strowger's automatic switching system.

The automatic dial system changed telephony forever. Almon Brown Strowger (pronounced STRO-jer) was born on 1839 in LaPorte, Indiana, the city in which he later installed the first commercial automatic exchange.

Strowger was an undertaker. His invention was developed because someone was stealing his business and he sought to do something about it. Telephone operators, perhaps in league with his competitors, were routing calls to other undertakers. These operators, supposedly, gave busy signals to customers calling Strowger or even disconnected their calls. Strowger thus invented a system to replace an operator from handling local calls.

Like Bell, Strowger filed his patent without having perfected a working invention, describing the switch in sufficient detail and with enough novel points for it to be granted Patent number 447,918, on March 10, 1891. Again like Bell, Almon Strowger lost interest in the device once he got it built. It fell upon his brother, Walter S. Strowger, and Joseph Harris, who helped promoted the switch and provided investment money.

Without Harris, soon to be the organizer and guiding force behind Automatic Electric, dial service might have taken decades longer for the Bell System to recognize and develop. Competition by A.E. forced the Bell System to play switching catchup, something they really only accomplished in the 1940s with the introduction of crossbar switching.

In 1893 the first central office exchange with a common battery for talking and signaling began operating in Lexington, Massachusetts.

This common battery arrangement provided electricity to all telephones controlled by the central office, where before a customer's telephone needed its own battery to provide power.

The common battery had many consequences, including changing the basic design of the telephone instrument. Big and bulky wall sets with their wet batteries providing power and cranks for signaling the operator could be replaced with sleek sets that fit on desks.

In 1899 American Bell Telephone Company reorganized yet once again. In a major change, American Bell Telephone Company conveyed all assets, with the exception of AT&T stock, to the New York state charted American Telephone and Telegraph Company. The rationale was that New York had less restrictive corporate laws than Massachusetts did.

In 1900 loading coils came into use. Patented by Professor Michael I. Pupin, loading coils helped improve long distance transmission. Spaced every three to six thousand feet, cable circuits were extended to three to four times their previous length.

Essentially a small electro-magnet, a loading coil or inductance coil strengthens the transmission line by decreasing attenuation, the normal loss of signal strength over distance. Wired into the transmission line, these electromagnetic loading coils keep signal strength up as easily as an electromagnet pulls a weight off the ground. But coils must be the right size and carefully spaced to avoid distortion and other transmission problems.

In 1901 the Automatic Electric Company was formed from Almon Strowger's original company. The only maker of dial telephone equipment at the time, Automatic Electric grew quickly.

The Bell System's Western Electric refused sell equipment to independents, so Automatic Electric makers like Kellog and Stromberg-Carlson found rapid acceptance. By 1903 independent telephones numbered 2,000,000, of which Bell managed 1,278,000.

Big, Bad Bell

Bell's reputation for high prices and poor service worsened with increased competition. When bankers got hold of the Bell System, it faltered.

In 1907 Theodore Vail returned to the AT&T as president, pressured by J.P. Morgan himself, who had financial control of the Bell System. Morgan was a true robber baron and thought he could turn the Bell System into America's only telephone company. He bought independents by the dozen, adding them to Bell's existing regional telephone companies. AT&T management finally organized the regional holding companies in 1911, a structure that held up over the next seventy years.

Morgan also worked on buying all of Western Union, acquiring 30% of its stock in 1909 then installing Vail as its president. Vail thought telephone service was a natural monopoly, much as gas or electric service, but he also knew times were changing and that the present system couldn't continue.

In January 1913 the Justice Department informed the Bell System that the company was close to violating the Sherman Antitrust Act. Things were going badly with the government, especially since the Interstate Commerce Commission had been looking into AT&T acquisitions since 1910.

J.P. Morgan died in March 1913; Vail lost a good ally and the strongest Bell system monopoly advocate. In a radical but visionary move, Vail cut his losses with a bold plan. On December 19, 1913, AT&T agreed to rid itself of Western Union stock, buy no more independent telephone companies without government approval and to connect to the independents with AT&T long distance lines.

Rather than let the government remake the Bell System, Vail did the job himself. Since the independents paid a fee for each long distance call placed on the Bell network, and because the threat of governmental control had eased, the Bell System grew to be a de facto monopoly within the areas it controlled, accomplishing by craft what force could not do.

To this day, 1,435 independent telephone companies still exist, often serving rural areas the Bell System ignored (and still ignores).

Long Distance, Thanks to the Electron Tube

In 1906 Lee de Forest invented the electron tube. Its amplifying properties led the way to national phone service. Long distance service was still limited. Loading coils helped to a point but no further. Transcontinental phone traffic wasn't possible, consequently, a national network was beyond reach.

In 1907 Theodore Vail instructed AT&T's research staff to build an electronic amplifier based on their own findings and De Forest's pioneering work. They made some progress but not as much as de Forest did on his own. AT&T eventually bought his patent rights to use the tube as a telephone amplifier. Only after this and a year of inspecting De Forest's equipment did the Bell Telephone Laboratory make the triode, an amplifying electron tube, work for telephony.

The triode in particular and vacuum tubes in general would make possible radiotelephony, microwave transmission, radar, television, and hundreds of other technologies. Telephone repeaters could now span the country, enabling a nationwide telephone system, fulfilling Alexander Graham Bell's 1878 vision.

The vacuum tube repeater ushered in the electronics age. The device was a true amplifier, powered by an external source, capable of boosting strength as high as was needed.

As evidence of the triode's success, on January 25, 1915 the first transcontinental telephone line opened between New York City and San Francisco. The previous long distance limit was New York to Denver, and only then with some serious shouting.

Two metallic circuits made up the line; it used 2,500 tons of hard-drawn copper wire, 130,000 poles and countless loading coils. Three vacuum tube repeaters along the way boosted the signal. It was the world's longest telephone line. In a grand ceremony, 68 year old Alexander Graham Bell in New York City made the ceremonial first call to his old friend Thomas Watson in San Francisco.

In 1921 the Bell System introduced the first commercial panel switch. Developed over eight years, it was AT&T's response to the step by step switch. It offered many innovations and many problems. Although customers could dial out themselves, the number of parts and its operating method made it noisy for callers. The switch used selectors to connect calls, these mechanical arms moving up and down in large banks of contacts.

In 1934 a New Deal measure enacted the Federal Communications Commission. The FCC began investigating AT&T and every other telephone company, issuing a 'Proposed Report' after four years

The commissioner denounced AT&T for unjustifiable prices on basic phone service and urged the government to regulate prices the Bell System paid Western Electric for equipment, even suggesting that AT&T should let other companies bid for Western Electric work.

At that time AT&T controlled 83% of United States telephones, 91% of telephone plant and 98% of long distance lines. The outbreak of World War II, two and a half months after the final report was issued in 1939, staved off closer government scrutiny.

In 1936 coaxial cable was installed between New York City and Philadelphia. In 1937 the first commercial messages were sent through it. Multiplexing developed, letting toll circuits carry several calls over one cable simultaneously.

By the mid 1950s, 79% of Bell's inter-city trunks were multiplexed. Multiplexed signals eventually moved into the local network, improving to the point where one circuit could carry 13,000 channels at once.

One major accomplishment was directly related to WWII. Fearing its radio and submarine cable communications to Alaska might be intercepted by the Japanese, the United States built the longest open wire communication line in the world began operating between Edmonton, Alberta and Fairbanks, Alaska. Built alongside the newly constructed Alcan Highway, the line was 1500 miles long, used 95,000 poles and featured 23 manned repeater stations.

In 1938 the Bell System introduced crossbar switching to the central office, an improvement on work done by a Swedish engineer, Gotthilf Ansgarius Betulander.

Installed by the hundreds in medium to large cities, crossbar technology advanced in development and popularity until 1978, when over 28 million Bell system lines were connected to them. (Panel switching lines peaked in 1958 at 3,838,000 and step by step lines peaked in 1973 at 24,440,000.)

On June 30, 1948 the Bell System unveiled the transistor, revolutionizing every aspect of the telephone industry and all of communications.

Capitalizing on a flowing stream of electrons, along with the special characteristics of silicon and germanium, the transistor dependably amplified signals with little power with little heat. Equipment size was reduced and reliability increased.

In August, 1951 the first transcontinental microwave system began operating. One hundred and seven relay stations spaced about 30 miles apart formed a link from New York to San Francisco. It cost the Bell System $40,000,000.

In 1954 over 400 microwave stations were scattered across the country. By 1958 microwave carrier made up 13,000,000 miles of telephone circuits or one quarter of the United States' long distance lines.

Microwave wasn't possible over the ocean and radiotelephony was limited. Years of development lead up to 1956 when the first transatlantic telephone cable system started carrying calls. It cost $42 million. Two coaxial cables about 20 miles apart carried 36 two-way circuits. Nearly 50 sophisticated repeaters were spaced from 10 to forty miles along the way. Each vacuum tube repeater contained 5,000 parts and cost almost $100,000. The first day this system took 588 calls, 75% more than the previous 10 days average with AT&T's transatlantic radio-telephone service.

In the mid-50s Bell Labs launched the Essex research project. It concentrated on developing computer controlled switching, based upon using the transistor. It bore first fruit in November, 1963 with the 101 ESS, a PBX or office telephone switch that was partly digital.

In 1956 AT&T agreed under government pressure not to expand their business beyond telephones and transmitting information. Bell Laboratories and Western Electric would not enter the computer and business machines industries. In return, the Bell System was left intact with a reprieve from anti-monopoly scrutiny for a few years.

Recent History

The 1960s began a dizzying age of projects, improvements and introductions. In 1961 the Bell System started work on a classic cold war project, finally completed in 1965. It was the first coast to coast atomic bomb blast resistant cable. Intended to survive where the national microwave system might fail, the project buried 2500 reels of coaxial cable in a 4,000 mile long trench. 9300 circuits were helped along by 950 buried concrete repeater stations. Stretched along the 19-state route were 11 manned test centers, buried 50 feet below ground, complete with air filtration, living quarters and food and water.

In 1963, digital carrier techniques were introduced. Previous multiplexing schemes used analog transmission, carrying different channels separated by frequency. T-1 or Transmission One, by comparison, reduced analog voice traffic to a series of electrical plots, binary coordinates to represent sound. T-1 quickly became the backbone of long distance toll service and then the primary handler of local transmission between central offices.

In 1965 the first commercial communications satellite was launched, providing 240 two-way telephone circuits, and the 1A1 pay phone was introduced by Bell Labs and Western Electric after seven years of development. Replacing the standard three-slot pay phone design, the 1A1 single slot model was the first major change in coin phones since the 1920s.

1965 also marked the debut of the No. 1ESS, the Bell Systems first central office computerized switch. The product of at least 10 years of planning, 4,000 person-years of research and development, and $500 million, the first Electronic Switching System was installed in Succasunna, N.J.

Built by Western Electric, the 1ESS used 160,000 diodes, 55,000 transistors and 226,000 resistors, capacitor and other components. These were mounted on thousands of circuit boards. Not a true digital switch, the 1ESS did feature Stored Program Control, a fancy Bell System name for memory, enabling all sorts of new features like speed dialing and call forwarding. Without memory a switch could not perform these functions; previous switches such as crossbar and step by step worked in real time, with each step executed as it signaled.

In 1968 Carter Electronics Co. of Dallas, Texas challenged the telephone equipment monopoly in court. Carter provided a simple device called the Carterfone to connect mobile radio users to the telephone network. The device was nothing technically difficult; amateur radio operators had used home-made phone patches for years.

Southwestern Bell took Carterfone to court and lost. The Carterfone case was the first chink in the telephone companies' armor. The FCC established rules permitting connection to the local network. The Bell System argued that devices had to be manufactured to tight specifications to prevent harms to the network and its maintenance personnel. The damage possible was purportedly high voltage levels that could crosstalk into other circuits, and harmful voltages and currents that could damage equipment or shock technicians.

The solution that the FCC accepted was interconnection through a protective coupling arrangement (PCA). PCAs were provided by the phone company at a cost, and provided an interface to which customers could connect their equipment. Customers were inconvenienced, but were not restricted to Bell System-owned devices.

This was the birth of the interconnect industry.

Later, the PCA requirement was eliminated and in its place a registration process supervised by the FCC was instituted. Only registered devices were (and still are) permitted to be connected to the network. This provision includes devices manufactured and owned by the ILECs.

At the end of the 1960s AT&T began experiencing severe customer service problems, especially in New York City having to do with unforeseen demand coupled with reduced maintenance. The Bell System fixed the problems, but not without an attitude that embittered people by the millions.

Bell was not alone in dealing with dissatisfied customers. GTE also had problems. GTE and Automatic Electric went through tremendous growth in the 1960s. Automatic Electric expanded to four different facilities and cut over their first computerized switch in Saint Petersburg, Florida in September 1972. It was called the No. 1 EAX (Electronic Automatic Exchange).

In 1969 Microwave Communications International (MCI) began transmitting business calls over their own private network between Saint Louis and Chicago. By bypassing Bell System lines MCI offered cheaper prices.

AT&T strenuously opposed this specialized common carrier service, protesting that Bell System's long distance rates were higher since they subsidized local phone service around the country. Still, MCI was a minor threat, economically. The real problem started a few years later when MCI tried to connect to the Bell System network.

MCI's customers, like Carter's, began asking to connect to the local networks in both cities. MCI accommodated them with a service they called Execunet. The MCI switch was connected between its microwave network and local telephone company trunks. Users could dial into the MCI switch, identify themselves with a PIN number, dial the destination telephone number, and MCI completed the call through the telephone company's networks.

Again, AT&T objected, arguing that MCI's service was in direct competition with its monopoly. Again, the courts ruled in favor of.

In 1975 the Department of Justice (DOJ) filed an anti-trust suit against the Bell System.

The 1975 anti-trust case was occasioned by numerous complaints from other companies about the AT&T monopoly. MCI and other carriers were vocal in claiming (accurately) that their connections were technically inferior to those AT&T provided to its long distance customers. The telephone companies' customers could dial 1 for long distance and be automatically identified, but other common carriers' customers had to dial a PIN number. Transmission was inherently better on the AT&T network, and callers automatically received AT&T long distance unless they dialed a local telephone number to bypass AT&T.

Competing manufacturers complained that the Bell markets were closed to all but the AT&T captive supplier, Western Electric. The Bell operating companies (BOCs), suppliers argued, automatically selected Western Electric equipment even though competitors' equipment might be technically superior.

The Department of Justice was asking to dismantle the Bell System.

AT&T argued that the structure of the Bell System operated in consumers' interests, and that regulation was an OK substitute for competition. They argued that the BOCs were required by law to subsidize local service rates through a complicated division of revenue procedure, and that competitive carriers were not saddled with this burden.

Experts argued that Bell Laboratories was a source of research and development that the nation could not possibly afford to lose. Others said out the nation's telephone network was the best in the world, and that dismantling it would not be in the public's interest. Other experts argued the other side: that technical progress was being impeded by the lack of competition, and that prices were kept artificially high since the Bell System had no incentive to economize.

In the midst of the arguments in the DOJ case, the surprise announcement came that AT&T had agreed to dismantle itself. The parent company would keep its long distance business and its manufacturing arm, Western Electric, and would be free to sell customer premises equipment.

The BOCs would be spun off into seven independent regions. They would be required to provide equal access to the local exchange to all long distance providers. They were prohibited from manufacturing equipment, and they were prohibited from offering service across artificial local boundaries known as LATAs. Finally, their procurement practices would be subject to open competition.

Bell Laboratories was divided into two segments. One was fundamental research and operations-oriented research on behalf of the BOCs. The other was product-oriented development for Western Electric. Accordingly, BOCs and Western Electric shared funding. The manufacturing-related portion of Bell Labs stayed with AT&T under divestiture, and the remainder was formed into Bell Communications Research (BellCore). The seven RBOCs jointly owned BellCore.

The Modified Final Judgement (MFJ) remained in force until the Telecommunications Act of 1996. The Act of 1996 modifies certain of the restrictions on the incumbent local exchange companies (ILECs) and imposes others. Over the next several years the nation's telecommunications network will be fractured and restructured. Instead of being driven by regulation, it will be driven by competition.

The Telecommunications Act of 1996, The Local Loop, and Competitive Local Exchange Carriers

The Telecom Act of 1996 was enacted by Congress in February, 1996 "to provide for a pro-competitive, de-regulatory national policy framework designed to accelerate rapidly private sector deployment of advanced tele-communications and information technologies and services to all Americans by opening all telecommunications markets to competition."

The Telecommunications Act of 1996 requires that ILECs (Incumbent Local Exchange Carriers), also called RBOCs (Regional Bell Operating Companies), open up their local telephone markets to competition.

The RBOCs are federally mandated to allow their competitors to interconnect with their local connections and wires by unbundling their networks and/or by reselling their network elements.

Any company can now offer services that require access to businesses and homes over local telephone connections. These new local service competitors are referred to as CLEC, Competitive Local Exchange Carriers. A CLEC offers its services to its customers by connecting the ILECs' "unbundled" network elements to its own switches and other systems.

Title I of the Act forces ILECs to open up their Central Office (CO) facilities to competitive service providers in exchange for entry into the lucrative long-distance and equipment manufacturing markets. As long as an ILEC is not offering CLECs access to existing facilities and networks, they aren't allowed to enter the long distance market.

Mutual Compensation, UNE (Unbundled Network Elements), Colocation and Resale

Broadly speaking, the Act gives CLECs three ways to compete with the ILECs: (1) network-to-network interconnection; (2) unbundled elements; and (3) resale.

<B>Mutual Compensation</B>

Network-to-network connections require payment from the company using the facilities to the company that owns the facilities. Most of the time the carriers pay each other because they're accessing each others' facilities or equipment every time a call is handed off.

The FCC calls these fees that interconnecting carriers pay to terminate traffic on each other's networks "mutual compensation."

The payments for mutual compensation are typically between two and four cents per minute as established by FCC rules. When the traffic is intra-state or local, the state commissions are charged with implementing and establishing mutual compensation payments. Long distance mutual compensation is handled by the FCC.

Unbundled Network Elements (UNE) and Colocation

The Telecom Act of 1996 makes sure that CLECs have access to rented floor space in the ILEC Central Offices (COs) and to any of the circuits terminated in that CO. Access to the floor space is known as colocation, and the access circuits are known as local loops. Local loops connect each residential or business subscriber in a given area to its CO.

A Section 251(c)(3) of the Act says that ILECs (Incumbent Local Exchange Carriers) must make "unbundled network elements" available to any competing telecommunications carrier for any communications service.

The Act defines an unbundled network element (UNE, pronounced "you-knee") as any "facility or equipment used in the provision of a telecommunications service," as well as "features, functions, and capabilities that are provided by means of such facility or equipment."

The FCC has identified a minimum list of seven UNEs that ILECs must offer, including such items as switches and inter-switch transport of telecommunications traffic. Practically, most CLECs want as little to do with ILECs as possible, so CLECs are most concerned about obtaining access to the portion of the ILEC network that it is most difficult and expensive to duplicate, the local loop.

The "local loop" (also known as the "last mile") connects the ILEC Central Office switches to their customers. The loop is one of the minimum UNEs identified by the FCC.

The Local Loop

The "local loop" or "last mile" refers to the last piece of (what's usually copper) wire that connects your building to the nearest Central Office (CO).

The Telecommunications Act of 1996, in part, deregulated the local loop. Now CLECs (Competitive Local Exchange Carriers) can use these wires with their equipment and sell new services like high-speed digital data.

The advantage the ILECs have over the new competition, at least for now, is that they own the copper cable loops to virtually every business and residence in the country. If CLECs want to reach residences and small businesses, they usually have to use, and pay for, ILEC loop facilities.

Copper local loops will probably not be replaced with fiber optic service any time soon. Converting to fiber optics at the residential and small business level is not yet worth the investment that is required: carriers estimate that that bringing fiber optic to the home will cost about $1,000 per residence.

Recently, the local loop has been used as is to push digital services. ISDN, xDSL, and Frame Relay services are being marketed by CLECs and ILECs both. These high-speed digital transmission schemes use the existing copper and can deliver video, telephone, Internet access, plus lower bandwidth services like alarm reporting and meter reading on the same old copper local loop.

> **See the Provisioning and Transport Services section of this book for more information about digital services.**

Local subscriber loops are connected to the central office switching equipment using hardware called DLCs (Digital Loop Carriers). A DLC is T-1 carrier interface designed specifically for subscriber loops. DLCs connect to the subscriber lines in the central office, route the signals over a T-1 channel, and convert them back to a subscriber line signal at the remote end.

Colocation
Section 251(c)(6) of the Act says that the ILECs have to allow their competitors to "collocate" their equipment on Central Office premises, so that they can connect their facilities to the central offices' UNEs.

Using a colocation strategy, a CLEC will buy one or more switches and install fiber optic facilities connecting its switch(s) to one or more of the ILEC's switching centers. The CLEC obtains local loop UNEs from the ILEC to connect the switching centers to individual customers. The CLEC colocates the equipment need-ed to connect those loops to their own fiber, and provides service to the customers using their own switch, their own inter-switch fiber, and the ILEC's local loops.

Colocation is the only connection that wireless carriers (cellular or PCS providers) and cable companies offering telephone service need to connect to the ILEC's network.

Wireless providers, cable companies and bigger "wired" CLECs have stand-alone networks, with their own switches, and connec-tions from the switch to the customer. All that these companies need to be in business is a connection that enables their customers to call the ILEC's customers over the PSTN and vice versa.

Resale

The third way for CLECs to enter the local market is provided by Section 251(c)(4) of the Act and requires ILECs to offer any telecommunications service they offer to their retail customers also to competing providers, but at a wholesale discount.

The discount is calculated by subtracting from the retail price the amount of money that the ILEC incurs in performing "retail" functions such as marketing, sales, and billing.

A facilities based reseller builds its own network and provides switching services within the PSTN or using equipment that is colocated at Central Offices.

CLECs that resell local or long distance services don't necessari-ly have to invest in switches, fiber optic transmission facilities, or colocation arrangements. All of the underlying capital investment can be provided by the ILEC, and the reseller competes for your business based only on price.

Setting an ILEC's actual wholesale discounts is a very complex and contentious process that the Act delegates to the state public utilities commissions. Issues relating to wholesale discounts are presently being litigated, but in most cases the discount will not be higher than 25%.

This is a slim margin for resellers who still must cover marketing, billing, general overhead, bad debt and other costs of doing business.

Regulatory Reality

The states are charged with implementing interconnection under the Act. A company that desires to interconnect with an ILEC must become a certified CLEC (Competitive Local Exchange Carrier) and negotiate an interconnection agreement with the telephone company.

If the CLEC is offering only intrastate services, it need only apply for carrier certification at the state level. If it is offering interstate services, the CLEC must be certified by the FCC too.

The RBOCs are successfully challenging the specifics of the Act in the courts and portions of it have been overturned. Most recently, the very constitutionality of any restrictions on the Baby Bells' entry into long distance was successfully challenged in court, and the restrictions were overturned by the court, although they are temporarily stayed on appeal.

Although the stated purpose of the Act was to foster competition, Title V (Sec. 509) made it federal policy to promote "the Internet and other interactive computer services and other interactive media." Accordingly, use of the Internet was exempt from the paying access charges to the local phone companies.

This exempted Internet Service Providers from paying access charges to the ILEC when they use the Internet to transport voice calls. The exemption has also given rise to a whole new computer telephony technology, Internet Telephony (or VoIP, Voice Over the Internet Protocol).

See the chapter called *Voice over the Internet or Intranets* - for more about telephony over the Internet.

Telecommunications Subcontractors and Consultants

It is difficult to assemble, train and maintain a staff to manage all aspects of your telecommunications systems and services. The solution may be to look outward to other firms or individuals for help.

Types Of Management Services

Almost any type of activity for managing telecommunications can be handled with contract help, including an office relocation, telephone system acquisition, office expansion, PBX upgrade or telephony technological additions like voicemail or Computer Telephony.
Some other telecom-related services that subcontractors or consultants can offer are:

Ongoing record keeping, either manual or computer-based.

Making program changes to the telephone system or related systems such as voicemail.

Placement and tracking of telephone system work orders.

Administration of telecom expenses and cost allocation systems, including the review and verification of your monthly telephone bills.

Preparation and maintenance of corporate telecommunications budgets and corporate directories.

Liaison with telecom vendors on a day-to-day basis.

Evaluation of competitive services offered by local and long distance carriers.

Training of users, switchboard attendants or system administration.

How to Find and Manage Outside Support

The best way to find help is to ask around. Talk to your associates with similar responsibilities at other firms to see whom they have used and what their experience has been.

Talk to members of user groups for the specific product or service with which you need help. There are brand-based PBX user groups that meet once or twice yearly.

Consulting is a word-of-mouth business. You can get a list of telecommunications consultants from the Society of Telecommunications Consultants (800-STC-7670). Ask your other vendors. Most are plugged into the local marketplace and can point you in the right direction if they can't offer the kind of help you need.

Don't stop at one referral, though, ask a few people. When you hear a name more than once, you know you have found someone good.

Once you have two or three names, check references. Checking references is key. Most outsourcers and consultants will include a general reference list, or perhaps a more detailed client reference list in the materials they give you. If the packet of information you get doesn't have a reference list, ask for one.

Ask about the reputation of the firm you are hiring. How long has they been in business? How many people do they have? What training have they had?

Define your needs carefully and try to get a person or service closely matching what you need. Make sure that the person or company who will provide the service has up-to-date and relevant knowledge and experience.

When you're discussing support services, make sure that you and the company you are talking to agree on the terminology. Here are some terms to describe the different levels of subcontracting or outsourcing services available.

Technician
A technician is typically someone who can do the things a purely administrative person cannot. Technicians usually handle physical changes, like connecting jacks for new telephones or installing new expansion boards in your telephone system.

Subcontractor or Independent Contractor
These terms are used interchangeably and typically refer to an individual who is providing work to you for a fee, but is not on your payroll. The fee is usually per hour or per day (per diem). Sometimes an entire firm is referred to as a subcontractor.

Complete Outsourcer
If you completely outsource your organization's telephone system and services, this means that you have delegated all responsibility to an outside firm, including the repair and maintenance of your PBX, provision of local and long distance services and review and payment of telephone bills. The outsourcer is paid a month-

ly fee, but may also make money reselling equipment and local and long distance service to you.

Telecommunications Management Outsourcer

This type of outsourcer manages your telephone system and service providers, but does not sell you equipment, or local/long distance service as a complete outsourcer may. These companies are the equivalent of having an in-house telecommunications department, but an outside firm staffs the "office."

Out-tasker

An out-tasker, a firm or an individual, takes complete responsibility for one or more specific jobs. For example, you may out-task the updating and production of your telephone directory or out-task the work order processing and record keeping for changes to your telephone system.

Consultant

A consultant is an individual who is charged with analyzing a problem or assessing a situation and providing a recommendation. A consultant may also execute steps necessary to carry out the recommendations.

When to Use a Consultant

A telcom consultant's job is to gather, assemble, and analyze information, then use it to advise and guide the client toward the products that best fit the client's needs.

Consultants are frequently called in when major purchases need to be made like new phone systems, voicemail, or other core components of telcom infrastructure.

Not everyone needs a consultant. If you know exactly what you want, or have a predisposition toward a particular brand or just have an obligation to get a couple of additional quotes, don't bother soliciting the help of a consultant.

Similarly, if you are looking at a smaller PBX or key system and you ren't going to use voicemail, ACD, IVR or any other complicated technology you won't need a consultant's help.

Big decisions and investments might require the experience and skill of a consultant. Here are a few things to consider a consultant for:

New facility

A new facility often creates logistical challenges that can be easily identified and addressed by a consultant. How will you run cabling for the phones, especially if you have warehouse space? Do you need a paging system? How will speakers be mounted? There are lots considerations and engineering issues to be ironed out before soliciting quotes. A consultant will have checklists, rules-of-thumb, and product information on file so you don't miss out on cost-effective solutions.

Sales environment

Have you gauged your potential inbound call traffic? How will all those calls get answered? How will you track your telemarketers' progress and plan your hardware and dialtone resources? There are a number of ways to collect data and solve these problems, and your consultant can advise you what you'll get out of each solution.

Big picture

Buying a new phone system isn't something you do every day. The industry has changed quickly in the last several years. If your knowledge isn't up-to-date or you have limited telecom experience, a consultant can provide good, basic, objective information. This education process might be best left to the consultant, an impartial third party, who is there to look after your best interests.

New network

Network integration is a paramount reason for changes in telecom hardware. If you're not clear on how network/telephony connections work, a consultant's help will be valuable in selecting a well-engineered group of products, and coordinating with multiple vendors to integrate and certify the solution you've purchased.

Working With a Consultant: The Buying Process

A consultant makes recommendations based upon your present telecom setup, your future growth plans, and how your business works. Standards and requirements are identified as part of this process, and a list of required features and options is created.

Consultants use information from you and their experience and knowledge to reate an RFP (Request For Proposal), a universal bid document to which prospective vendors respond. The RFP includes vendor information, provides requirements for guarantees with regard to service response, requests system and component pricing, and specifies other information that is required from potential vendors.

The RFP format is designed to help you compare vendors fairly by organizing bid information in a clear format. By becoming acquainted with your business, the consultant is also responsible for notifying vendors of any wiring requirements, and will act as an intermediary between you and the vendor.

Once the consultant creates the RFP document, and the vendors have responded, your consultant will evaluate the responses.

The consultant should narrow the field of responses, then schedule product demonstrations for you. System demonstrations let you see and feel the systems you're considering. The demo also provides an opportunity for you to understand how your needs will be addressed by each vendor. One or two vendors will probably emerge from the demo process as clear leaders.

After the demonstrations, you and the consultant will discuss what you've seen, select the best solution, finalize details, and award the bid. Then the consultant and your new vendor will take over, ordering the equipment and preparing for your installation. If you're buying advanced applications such as ACD, call accounting, network integration, etc., these will be engineered by your consultant and vendor (with your input, of course), prior to installation.

Finally, the system is installed and programmed. Phones are marked, placed on desks, and your users are trained. A good consultant will maintain a checklist (or fixlist) throughout this process and will make sure the vendor addresses them. After all the final details have been taken care of and you and the consultant are satisfied, your installation is complete.

For the Best Results

Think about the dynamics of in-house and outside support staff working together. Keep the lines of communications open between the groups. It's important to delegate responsibilities clearly so the in-house staff don't overlap with the outside support people.

Whether outsourcing, out-tasking or hiring subcontractors for daily help, ask for a work plan in advance, a detailed time sheet as the work is completed and verification that the work is complete.

Be clear about fees and responsibility for insurance, workmen's comp and payroll taxes. There are many and changing rules regarding the hiring of subcontractors.

Choosing A Dealer

You've done your homework and have a couple of phone systems in mind that fit your budget and have the features you've decided are crucial. Where do you look to buy?

Since buying a phone system puts you in partnership with the company you buy it from, shopping for a dealer should get the highest priority in your decision-making. Several dealers in your area probably sell the system you're looking for.

Ask other businesspeople in your area for phone companies they've used. Local dealers are always listed in the yellow pages, and bigger companies are on the Internet. You can also contact the manufacturer's corporate offices for a list of certified dealers near you.

In the last few years, the telephone dealer industry has changed greatly. Interconnect companies, that used to provide phone systems, installation, wiring and service, are now also selling and supporting PC-based peripherals and communications systems. Today's top dealers are able to configure, sell, install and support networking solutions that also happen to incorporate telephone systems.

Today's best dealers are versatile, and if they've survived the last five low-margin years, they're in for the long haul.

Bill Moretti is the Operations Manager of Automated Answering Systems (212-947-4155), one of New York City's best Lucent dealers. He attributes the company's success to its highly skilled and dedicated staff (sales, installation, maintenance and customer service). When a system is sold, all of the departments in their company work together to make sure that the system not only performs properly, but it also meets the customers' needs. After installation, if a customer calls for repair service, a move or change, or simply to ask a feature question, everyone at Automated takes great pains to make sure the customer is satisfied. "We treat our customers like gold," he says.

Automated answering systems has been around, as their name implies, since the dawn of automated telecommunications systems. Dedication to their customers has worked for them, and should be the philosophy you look for in a dealer local to you.

Bill suggests you look for the following characteristics when you're picking a dealer:

Certification

Make sure that the dealer is authorized to sell the system you're looking for. Manufacturer Authorization is not simply a matter of a dealer promising to sell a certain number of systems a year. Authorized dealers are carefully evaluated for size, financial stability, technical expertise, business ethics, etc.

Certified dealers are often trained by the manufacturer on an ongoing basis. An authorized dealer has priority access to inventory, manufacturer technical support and programming and training documentation; they'll be able to serve you better and resolve your problems faster.

Relationship

You want a dealer who's willing to develop a consultative, long-term relationship with you and your company. Favor any dealer who will assign a permanent, single-point-of-contact representative to your account, and make sure this person is not so overburdened with other accounts that they forget you exist between service calls.

Communication

Your dealer should be willing and able to speak with you on the technical level that makes you comfortable. Don't let the dealer believe you know more than you do because this will lead to confusion and time-wasting, and can cost money.

Support

A good dealer should be willing to work with the way your company makes buying decisions by reducing the demands made on you. If you have to justify buying decisions to a boss or committee, the dealer should provide you with helpful materials, and make him/herself available for further questioning, by higher-ups.

Your dealer should learn how your business operates and be there when you need help.

The dealer should always be sensitive to your need for minimizing cash outlay, where possible. But a dealer shouldn't let you nickel-and-dime yourself into a corner, either.

Versatility

Favor a dealer capable of providing turnkey service, who can handle (or efficiently and reliably subcontract) maintenance/programming, wiring, telco orders and installation supervision, resale of

LD or local service, wireless contracts and certifications, voice-mail and standard peripherals like call accounting or paging. It's always easier and often cheaper to deal with a single vendor that handles all of your communications needs

It's important that you find dealer willing to intercede with telephone service providers on your behalf. Especially for a new PBX, it's important for the dealer/installer and telco to work in sync.

How broad is the dealer's expertise in computing and data networking? Most telephony products integrate with data networks or require LAN services to deliver computer telephony benefits. Data-oriented telephony technologies can mean big savings and productivity benefits.

Recordkeeping

Your dealer should keep comprehensive client records. The dealer should know everything about your system, and have it written down. They should keep a record of your system programming, understand your trunks and telco services (including contracts, discounts and repair history), and have a complete and up-to-date physical map of your premises wiring.

Training

Find out what kind of training prospective dealers offer when they sell you a new phone system. Is it free? Are you allotted a specific amount of time?

Service

What kind of turnaround time does the dealer promise on service calls? Most offer 24-hour response for minor trouble and two or four hour response for major (more than 20% of stations or trunks dead) alarms.

Inventory

The dealer should keep a generous inventory (in the office and on trucks), so that they have the part you need, when you need it. Is the dealer willing to maintain a special inventory of items important to your business so that you can get new ones promptly when you need them?

You'll know when you find the dealer you're looking for. The company comes well recommended. The representatives are competent and give you a warm, fuzzy feeling. They sell and support the telephone system and other products you're looking for.

Communications Systems and Hardware

The Secondary Market

Many companies buy used or refurbished equipment to cut costs. This market is a great place to buy telephones and equipment if you take a little time to do your homework and browse for the best deals. Why not buy refurbished equipment if it can run as good as new, comes with a warranty and saves you money?

Secondary equipment usually costs 30% to 70% less than new. Most vendors will tell you that the older the equipment, the greater the discount. Whatever is scarce costs more money. Phone color also affects value; unpopular colors are scarcer, so cost more.

You can shop for the system you want through the manufacturer or dealer, then call a secondary vendor to see if you can get it for less. Make sure you know your product. The telephone and KSU model numbers indicate analog vs. digital, plus non-visible features like speakerphone.

If you decide to buy from the secondary market rather than a dealer or OEM (Original Equipment Manufacturer), make sure you can install it yourself or that you have someone lined up to install it. Secondary market sources are usually (but not always, so ask) equipment-only.

Many remarketers lease or rent equipment. They also assist in acquiring financing through third-party leasing companies or from the original equipment manufacturer (OEM).

Refurbished equipment is known for its reliability. Some say it's because it has already endured the burn-in that makes faulty circuits obvious. Refurbished products should be "near new" with all the OEM-provided accessories, up to date software and (at least) the standard warranty.

Make sure you get a user guide with each phone and programming manuals with the systems.

Many remarketers have increased their warranties from 90 days to up to two years. Find out if they provide advance replacement or if you must first return the defective goods. Find out who has to pay the shipping costs.

Remarketers focus heavily on the resale of parts. They should go out of their way to make sure they have in stock the parts you need. Some even track the average rate of failure for various components and stock their shelves according to the forecasted needs of their customers. Find out what the refurbisher's normal inventory level is for equipment. If they don't have well-stocked inventory, they may be less equipped to hook you up with the equipment you need when you need it.

Much of the market for secondary systems comes from companies that don't want to keep their own telecom inventory. Some vendors put serial barcodes on each item, so its history is always known.

Make sure you understand what you are buying. There are many terms that refer to different states of previously-owned and they are used somewhat interchangeably.

The National Association of Telecommunications Dealers

(NATD) defines refurbished equipment classes

As Is

Equipment that is bought or sold with no implied warranties. You should expect any condition from inoperative to good. This equipment may not be complete. Buy at your own risk.

Fair Condition Equipment

This equipment is usually in working condition but looks poor.

Good Condition Product

Equipment that is in working condition and looks good.

New

Generally defined as being sold by an authorized vendor of the Original Equipment Manufacturer (OEM) carrying the OEM's standard warranty.

Like New Excellent condition

Under normal conditions could pass as new, (not used) but is not necessarily in the Original Equipment Manufacturer (OEM) packaging.

Refurbished

Refurbished equipment is cleaned, repaired, and/or painted (panels, covers, etc.) to restore the appearance of the product to a like new condition. It is completely tested, repaired and ready for installation.

Factory Refurbished Equipment

Factory refurbished equipment has been returned to the factory and the factory has replaced the plastic, repaired what's broken, upgraded circuit boards, or has otherwise reconditioned the equipment to near-new.

Reconditioned

Reconditionedis not a NATD term but it is usually used as synonymous with refurbished.

Before you buy, get a credit report, call other dealers and industry peers. Call the NATD (National Association of Telecommunications Dealers, 561-266-9440) to see if the dealer is a member. If so, see if they are in good standing.

Place several small orders to test dealers before placing a large order.

Refurbished Dealer Qualification Checklist

Ask these questions when shopping for a secondary market dealer:

How long has the company been in business?

What services are provided? Installation, repair, and adds, moves, and changes?

What is the level of expertise and support available? Ask about in-house techs and levels of knowledge, support after hours, time before callback.

How much and what kind of inventory is kept? How will the equipment be shipped to you? What is the inventory guarantee, if any?

What is the warranty policy? How long is it? Do they provide advanced replacement?

Ask what their definition of refurbished is. Are the telephone cases painted? Are the cords new? Is it the latest software version? How is the equipment tested?

What are their payment terms? Most secondary market vendors ship the first order COD and offer some kind of credit starting with the second order.

Are they a member of NATD? Ask and check their references.

Check how returns are handled. An RMA (Return Material Authorization) or RA (Return Authorization) is a code number issued to the buyer to faciliatate the return of a product for repair, replacement, or refund. It usually has to be on the outside of the box of the equipment being returned.

Also keep this info handy: date ordered, salesperson, shipping carrier, product name and model number, warranty terms.

The refurbished market is the best place to get additional phones and expansion cards, cheap. Buying entire systems from a remarketer will save you money, but unless you understand what you're buying and can install it or get it installed, you might have more trouble buying refurbished than not.

See the appendices of this book for a list of vendors of refurbished equipment.

How To Install A Telephone System

Designing, purchasing and installing a new phone system is a crucial project. Designate someone inhouse who is accountable for assessing your needs and has the time to make a well-researched recommendation.

Large phone system installations are almost never smooth. But the more organized you are and the more information you have to give to your vendor, the better the cutover will go.

The single most important part of installing a communications system is the planning. Discuss your communications needs internally, and decide what your company needs. Then discuss how your business operates with your consultant, dealer and installers.

You need to know what you want a communications system to do. Communicate with your users. Find out what your people need to do their jobs. Find out what features they actually use. What features would help them do even better work with their phones?

The more information you get from the people who will use the system, the better prepared everyone will be.

Call Handling and Design

How should calls be handled? How (and from whom) do calls flow within and between departments? How should they flow? Now's the time to match your system design to how your business should (or does) actually communicate.

Draw a picture of how your calls come into your business and show how they'll be switched under every possible condition. You'll need a call flow diagram for the phone system, the ACD, the autoattendant/voicemail and the IVR system. Decide how incoming calls will be answered, how many rings before calls are transferred to voicemail, who will back up whom, and so on.

Who will be calling into your organization, and what number will they be dialing? Do you have a dedicated receptionist who handles incoming calls or should you use DID (Direct Inward Dial) and let individuals handle their own messaging? (What's best is probably a combination of the two.)

When does which type of call go to voicemail? To the call center/ACD? To a backup person or operator? Will you have off-premises forwarding from extensions to pagers or cell phones?

Call Flow Design Hints

1. Think about where the calls are coming from to decide how they should be handled. Incoming calls to your toll-free number should be kept short. Don't have calls you're paying for spend a lot of time in IVR menus or on hold waiting for an agent.

2. Your telephone system and adjuncts should enhance your customer service, not harm it. Always have an option for callers to speak to a real person. You can set up a series of call handling routes that terminate at a real person, minimizing the amount of human time required.

3. A good way to set up an informal call backup system that does-n't require secretaries or a central answering system is a depart-mental hunt group. Two or three extensions back up each other, and then the call can forward to voice mail. Or vice versa; voice-mail first and a person as a last resort. (This setup assumes that someone in the group is covering the phones all of the time.)

4. Most systems have a delayed ringing feature. Delayed ring works as a good substitute for call forwarding from an unan-swered phone to a backup person, for instance, from executive phone to admin phone. The call can ring three or four times on the first phone, and then a button light of that extension starts ringing on another phone. The button on the backup phone blinks but doesn't sound for the first few rings.

5. Be aware that once a call forwards out of your premises system, it can't come back. If you use off-premises forwarding (to another landline-based number), or if you forward calls to a cell phone you'll need another voicemail box or answering machine for the number the calls are being forwarded to.

6. Get expert advice, or become an expert yourself, about how the interfaces between the phone system and your adjunct systems work. Digital connections often work most reliably, but can also be prohibitively expensive. It's important that the IVR/Auto Attendant and Auto Attendant/Voicemail systems be able to move calls, and any information like caller ID that travels with them, back and forth between each other.

7. It can make sense, (if you can afford it) to set up private exten-sions alongside extensions that handle your business calls. While you don't want people making and taking private calls at your expense, they do need to call home and make other life-organizing communications. Personal calls can wreak havoc on your business call design.

8. You need to know how each call will be handled from the moment it hits your business until the call is completed. Don't assume that once a call has been answered by voicemail the communication is finished and the caller is satisfied. It is important to make sure that your users are reading or listening to their messages and responding to them.

Telephone Permissions

There are two ways to control how your telephones are used. The Class of Service (COS) determines how and where calls are made from the extension(s). Button tables are a graphic representation, or template, of a telephone design.

Class of Service (COS)

An extension's COS is represented to the phone system by a (usually) two-digit number. Individual COS classes specify whether the extension with that COS can dial out of the building, locally, long distance, internationally or some combination of the above.

COS also determines how an outbound call from the extension is handled by the system. Executive callers can be given top priority in the outbound queue, so they never get blocked, for instance.

Button Tables

Button tables (also called feature tables) are usually represented by a stylized picture of the telephone's faceplate. Each type of telephone can have several button tables, depending upon the number of extensions that will appear on that phone and the features the phone will have.

Using a few standard telephone setups with the feature buttons all in the same place facilitates training and re-programming in the future. It also lets people move from phone to phone easily.

Hardware and Phone Features

You'll have to decide who gets what type of phone (executive sets with LCDs, non-LCD phones, analog sets, speakerphones) and what features users will have access to.

Make a list of all of your phones by user name, extension, instrument type, location, Class of Service (COS) and button assignments. List everyone who's going to need phones, and every place that has or needs phones (don't forget the warehouse and lunchroom).

An easy way to keep your telephone plan is a keysheet. (There's one in the back of this book to use as a template.) Keysheets are spreadsheet grids, with the user names on the vertical left side and each phone's button assignments along the top horizontal. Don't forget to identify fax and modem jacks.

Mark a floor plan that shows telephone, fax and modem locations, their extension numbers and purposes. This is extremely useful to installers, especially for systems with over 50 stations.

Provisioning in Brief

See the Provisioning chapter in this book for much more about how to choose trunks and services.

In the process of planning your system, you should identify all the lines coming into your office and see if you need more, less, or different types of trunks altogether. Contact your local phone company and request a copy of your Customer Service Record (CSR). This will tell you what numbers, trunks and features you're being billed for.

Have an installer check your trunks for what you have and what you are actually using, before you place an order for any new service. You might find lines you don't have that you're being charged for and lines you should have that you don't.

You now have the opportunity to choose a competitive local provider. What used to be called "one-plus" or local-long distance calling is now a competitive market. You can still use your local telephone company, hitch your local-long distance calling to your long distance bill, or get access, local and long distance service from three separate providers.

If you plan to continue using the services of your existing long-distance carrier, be sure to call that company directly and give them your new numbers in advance. Call them again as soon as your cutover takes place, and confirm that they are carrying your long-distance traffic, and that the features you subscribed to such as toll-free service or accounting codes carry over with these changes in your local telephone service.

If your company is moving to a new location, you won't be able to keep your existing telephone numbers unless the new location is within the same Central Office service area. If your move is within the service area, the telephone company can provide duplicate service, usually called "half-tapping" or "back-tapping."

Submit your telco (telephone company) order well in advance of your projected installation date. DID trunks can take up to eight weeks to install. Central Office trunks often take 10 business days. The telephone company can install numbers incorrectly or not working at all, so you'll need time for repair. Your vendor can place your telco order for you with your written permission called a letter of agency.

Cutover

Make sure that when there is a "cutover" (your phone system and telco trunks are switched over, usually with some service outage), that it is scheduled for a time when there will be minimal disruption to your business. Get the account executive or engineer to commit to a date and time window, and confirm it with you in writing.

Wiring

A new phone system installation should have new wiring. Now is a good time to rewire, with growth and new services in mind. In new offices, it is common to see each desk served by a multi-purpose phone jack, which has plugs for two separate phone cables, and for LAN (local area network) service. By rewiring, you can make your office wiring not only tidy, but expandable, with additional cable available for future use. Two four-pair category 5 cables to each location will give you eight pairs of wires, enough for a phone, a network connection and spares for future use. See the Cable and Wiring chapter of this book for much more information.

Training

Receptionists and busy secretaries should have at least a couple of hours of one-to-one training before the system goes live. Other phone users should be encouraged to attend one of several scheduled classes before the system is installed. The first morning the system is live, a trainer should be onsite to go from desk to desk, answering questions.

Least Cost Routing

Least Cost Routing (LCR) is a method for routing calls over different carriers to get the best rate on a call-by-call basis.

LCR systems range in complexity from simple area code/carrier tables to complex systems that consider the day of week, time of day, and destination telephone number when deciding how to switch the call.

True LCR is based a table of area codes and/or NNXs with corresponding carriers assigned to handle calls to those areas. Only the largest PBXs support true LCR. Implementing true LCR is labor-intensive and tedious because you have to assign each area code individually to a trunk group or carrier. Make sure that you instruct your vendor to program LCR for you when the system is installed, so that it is part of your purchase price. Implementing it post-cutover will be expensive. Once the system is installed, get a list of the trunks that were programmed into each group and use this list to subscribe phone lines to the appropriate carrier. You'll have to pay for reprogramming if you change carriers or plans.

If you business has 50 or fewer phones, you can use a semi-automatic LCR method that uses trunk pooling and access numbers.

You can create trunk pools or line pools on most medium-sized and larger phone systems. A line pool is a group of outside lines. You can access a group by pressing an access number before you dial an outbound call

Each trunk group can have a different access number, and each group can be designated to a type of outbound call. You might dial nine for local calls, seven for intrastate, eight for international, and so on.

Use more lines in the groups that will carry the most calls.

Key Systems

A key system is a business telephone system that assigns Central Office (CO) trunks to telephones (allowing for multiple line pickups), has an internal intercom system, and adds features to making phone calls that traditional single line sets usually don't have.

Because multiple lines mean that calls can be picked up from several places around the office, key systems have historically been preferred in small- or medium-sized businesses. Today, key systems are still generally smaller than a PBX system, but they're boasting big-system features.

Most key systems have a controller box or cabinet, called a Key Service Unit (KSU) and proprietary (can only be used with that manufacturer's systems) telephones.

The KSU holds the system's switching components, power, line and station cards, and sometimes voicemail, paging and other outside system interfaces. All of the outside lines and phones plug into the KSU.

To make an outside call from a key system phone, you choose a free trunk (usually by pressing a button on the phone), and dial. (Some systems can be programmed to automatically seize an available or preferred line when you lift the handset.) Depending on how the system is programmed and how trunks are assigned, incoming calls may all ring, not ring at all, or ring only at certain sets.

The last couple of years have seen the introduction of key systems offering automated attendants and voicemail, consoles and computer telephony interfaces. Many key systems also now offer DID (Direct Inward Dial), T-1 and ISDN interfaces.

The most important thing to consider when buying a key system is how big the system is going to be. How many lines and phones will you need? Don't buy a system that's smaller than you need (you'll spend less now, and possibly much more later, for growing room.

Key systems typically fall into four categories

SOHO
Small Office/ Home Office) systems are aimed at the single person and very-small-office market. They generally support up to four trunks and 16 phones.

Many SOHO systems are "KSU-less" — there's no Key Service Unit. They make up for their small capacity with more smarts built into the sets. SOHO systems often use daisy-chain (from phone to phone) wiring, to bring the trunks directly to the individual phones. DC power and station-to-station signaling can also be distributed from phone to phone.

KSUless systems may offer lots of features at a great price, but they can be tricky to install, depending on your wiring and facilities. (You typically need one pair of wires in your cabling for each line, for instance, and many residences only have quad or two-pair wiring installed.)

Features like music on hold, SMDR (Station Message Detail Recording) output for call accounting and other features you want may not be available. They're also not designed to expand beyond rather strict upper limits — so if your business is growing, keep looking.

Small Key Systems

Small key systems usually run from three lines and eight phones to eight lines and 16 phones. The phones are installed using direct KSU-to-phone wires, (called "home runs") rather than daisy-chain connections from phone to phone.

Small key systems often don't have ports available for analog connections, so they can't support devices like modems and fax. They do have more system-wide features than KSUless systems do, however, like phone-to-phone intercom, external paging capability and music on hold.

Small key systems are often not expandable (called "closed" systems), so if you're planning to have more than 16 phones in your office any time soon, a better bet is a medium-sized system that will expand as your business does.

Medium-sized Key Systems

Medium-sized systems serve from about four to 24 trunks, generally up to 48 phones, have a central KSU, and also employ "home run" (KSU directly to station) wiring to the phones.

An "open" KSU is upgradeable by plug-in cards. Open systems let you start with a system suitable for your immediate needs, and will expand as you grow. Open systems are also more likely to provide analog ports for voicemail, fax machines and modems.

Some medium-sized key systems come with integrated voice mail and auto attendant, can handle digital and analog phones and will accept special types of trunks, like DID, T-1 or ISDN.

Large Key Systems

The large systems start at around eight trunks and go up to 32 or more. They can usually handle over a hundred analog and digital phones.

Large key systems are usually card-cage-looking systems that can be scaled and upgraded many different ways, including for different trunk connections like DID, T-1 and ISDN. Many are fully-open computer telephony platforms.

A large key system is often undistinguishable from a small PBX except for the upper capacities for trunk and station ports. Fully-featured, digital key systems are the norm, rather than the exception, at the upper end.

Digital vs Analog Key Systems

Every key system, even the smallest, employs digital signaling inside its KSU. Fully digital systems encode voice and system commands and send them digitally over one pair of wires from the KSU to each telephone.

If the key system you're looking at requires more than one pair of wires from the KSU to the phone to work, it's not digital end-to-end. Analog systems separate dial tone signals from power, intercom and sometimes data, so they'll require two or more pairs of wires from the system to the phone.

All-digital systems use Pulse Code Modulation (PCM) to sample and encode analog voice signals into a digital bit stream. The digital bit stream, or binary code, is sent over the backplane of the phone system's KSU to its destination extension. Once the call hits the extension, the digital desktop phone turns the voice signal back into an analog signal you can understand.

Digital systems have several benefits.

1. Call processing usually works faster when voice signals are digitized. Systems that work like this can simultaneously handle other traffic besides voice. By making an analog voice signal into a digital signal, you can transmit it simultaneously with other digital signals (like data and fax).

2. When a voice signal is switched and transmitted end-to-end in a digital format it will usually comes through with less noise than if an analog signal is sent. An electrical signal loses strength over distance, and must be amplified. In analog transmission, everything is amplified, including the noise and static the signal collects along the way, whereas most noise-induced interference is filtered out when digital signals are broken down and re-assembled.

3. Digital systems are also more likely to interface to digital trunks, like T-1, than analog systems. If you want to use a T-1 with an analog system, you'll probably need another piece of equipment, called a channel bank, between your incoming T-1 trunk and your key system.

Analog systems are great for startups and other budget-conscious firms. Small companies can save lots of money, especially buying refurbished, as long as latent needs for high-capacity digital communications and computer telephony interfaces don't surface before the system is amortized.

> **See the Secondary Market chapter for more information about refurbished equipment.**

Most analog key systems cost between $150 and $300 per station. Digital systems run higher; $300 to $600 per station. Per station price represents the system unit (KSU) price plus the cost of the phones, divided by the number of phones. Cabling costs are usually not included in the per station price.

KEY SYSTEM FEATURES AND ADJUNCTS TO LOOK FOR

Basic features that most key systems will have include:
The ability to restrict telephones, (called call blocking), multiple classes of service, conferencing capability, call waiting at an extension, distinctive ringing, flexible intercept, music on hold, night service, outbound call queuing, power failure transfer, system and station speed dialing, external paging access, call park, and do not disturb.

Data Ports
The hotter systems are employing "analog over digital" multiplexing technology to provide virtual analog ports on digital phones over one pair of wires. Instead of either an outside line jacked to your desk or a separate analog extension off of your phone system, these ports are in or attached to your phone, ready to use for fax or data.

Fax Detection
Fax detection capabilities eliminate the need (and costs) for dedicated fax numbers. The phone system understands fax tones and automatically routes incoming faxes to the extension to which the fax machine is attached.

On-Hold Capability
There should be a standard RCA-type on-hold port on the outside of the key system's KSU. You'll be able to hook up any music source, or, better yet, play your own advertisements using a digital announcer. Be warned, however, that it is illegal to re-broadcast radio stations or other music you haven't licensed. Performers are entitled to royalties every time their songs are played.

Caller ID

Some key systems can show the identity of incoming callers on the phone LCD display. You have to have Caller-ID service coming in from the PSTN (Public Switched Telephone Network), and the key system has to be able to read and understand the Caller ID signaling it gets.

You can do neat things with Caller ID to your phones, like store a log of the last few callers and call them back with the touch of a button. Caller ID improves customer service since you know who's calling and can be better prepared to handle the calls.

Voicemail

A conventional phone system and separate PC-based voicemail each typically incorporate an independent CPU, RAM, mass storage, call processing hardware and other resources. A fully-integrated phone system/voicemail unit can exploit shared resources and a common maintenance/management interface, and will often fit inside the system's KSU.

If you're going to want to use an auto attendant and/or voicemail, find out what products work with the system. Ask about how the voicemail integrates with the system. Many voicemail systems are PC-based and attach to the KSU using analog ports. This works, but if the key system has digital, integrated voicemail, you'll lose fewer calls. Analog interfaces don't always reliably transfer calls.

Computer Telephony

Lots of key systems offer Telephone Application Programming Interface (TAPI) adjuncts that equip a digital telephone for first party call control. Your power users can have unified messaging at their desktops.

More about computer telephony in the Computer Telephony section of this book.

Least Cost Routing and Toll Restriction

Least Cost Routing (LCR) employs user-defined look-up tables to find the least expensive long distance carrier service to send a call over, based on time of day and the area code/exchange dialed. Toll restriction lets you lock out certain international codes, area codes, exchanges or phone numbers.

Directories

Also called speed dial lists, directory lists are centralized lists of names and numbers (usually scrollable or searchable) programmed into a phone system or into individual phones. The directories are used as reference for the console operators and at station sets for quick or abbreviated dialing. Directories can be system-wide (stored in the KSU) or personal (stored in a station phone). Some systems will automatically make directories of in-house extensions.

Message Lamp

Most phone systems will talk to a voicemail system to light a lamp on the phones when there are messages in the mailbox assigned to that extension.

Screen Messages

This feature lets you send brief, pre-programmed messages to other display phones in the system. It lets someone on the phone respond non-verbally to someone who is trying to transfer a call, and internal callers can leave alpha messages on an unattended phone's LCD.

DID and T-1 Capability

DID trunks are a specialized telco service and need specific interfaces to work with a phone system. If you decide to use a T-1, find out if, and how, the key system you're looking at works behind a T-1. Many larger key systems have added built-in support for T-1 lines. This makes life easy. You plug the T-1 directly into the KSU. No external channel banks are needed.

> **For more about trunks, see the Provisioning section of this book.**

Automatic Call Distribution (ACD)

Integrated ACD functions let your small business act bigger. You can program ACD software to distribute calls in a variety of ways so that an incoming call always gets answered.

> **For more about ACDs, check the *Call Center Technology* of this book.**

Wireless

Some systems are wireless-ready. Most vendors offer one wireless replacement handset/base unit for infrequent deployment. If you have a sprawling enterprise and need extended wireless, check to see if the key system you're buying supports these phones, or look at buying a completely wireless system.

Automatic Set Relocation

This capability can save you big bucks. Some systems assign the programming of a specific telephone programming to the jack, in other systems the programming follows the set. If you buy the first type of system and want to move an employee from one desk to another you'll have to change the wiring or reprogram the switch. If your system supports automatic set relocation, you can just unplug the set and move it to another jack.

SMDR Output

SMDR (Station Message Detail Recording) is a data stream that issues from phones systems on-the-fly as calls are made, and includes specific information about each call. Call accounting systems use the call data to generate management reports like extension usage reports (good for identifying phone abuse) and traffic reports, so that you can see how much usage your facilities from the telco are getting.

Consoles

Consoles aren't generally available for key systems. They require different permissions and instructions from the system than regular phones do. You can often get away with a multi-button set, but check to see if consoles are supported, and if they are, see what they'll do that regular keysets won't.

Tips For Buying A Key System

Model Names

If the model name of a key system includes a number, it usually denotes the capacity. For instance 616 means six lines and 16 phones, 824 means maximum eight lines and 24 phones, and so on.

Analog Phone Compatibility

Analog phones are also called single-line sets or 2500 sets. They don't pack too many features, but they take more abuse than station sets and work just fine. Find out if the phone system you're looking at supports analog ports. You'll also need analog ports for fax machines, answering machines and modems.

Ease of Use

Once a phone system is installed and running your system will still require occasional tweaking to add or change users, lines and call handling options.

Don't be too dependent on service calls. They take time, and can become very expensive. Being able to program your own system will save lots of money in MAC (Moves, Adds and Changes) labor charges later.

Get a demonstration on how the system is programmed. Make sure you get programming (as well as user) manuals with the system and study them.

If you're going to program with a display phone, make sure the system uses an intelligent, menu-driven interface that's easy to understand and guides you through the steps. Some systems do this quite well. Some don't.

Some of the systems can be programmed with an attached PC. A computer screen is much more informative than even the biggest station-set LCD, and the PC keyboard is better than a touch tone pad for entering directory names and other alphanumeric data. Most manufacturers use pull-down menus to guide you through the programming steps.

Phones
Don't forget about the one thing you'll be using most: the phone. Key system phones usually run from a basic single line set with intercom to a many-buttoned set with a LCD display and speaker. Some manufacturers make all of their phones speakerphones, some don't, so check.

Make sure the keys on the phones are easy to see and well-spaced for dialing. The handsets should be comfortable, and the set should sit at an angle so that you can read the LCD easily.

Phone features should be easy to use, too. Make sure basic things like transferring and conferencing are easy to do.

Pick the Best Dealer
Get references. Make sure the technicians are certified. Make sure the vendor keeps spare parts for your system in stock. Get a guarantee of response time for MAC, major (usually defined as greater than 20% of your lines or stations are down) and minor outages.

Buy Bigger
Always buy a system with more phones and lines than you need. Make sure you ask salespeople how their systems grow and what the difference in price is if you buy more capacity now or add

capacity later. Buy a system that can be expanded to (the term is "wired for") at least 50% more than your current company size. Buy enough capacity, now, in terms of available ports, (the term is "equipped for") so that you can expand 20% without having to purchase expansion cards.

> **See the product and manufacturer lists at the end of this section for specific information about key system manufacturers and their equipment.**

Private Branch Exhange (PBX)

A PBX (Private Branch Exchange) is a high-capacity telephone system owned privately by the company that uses it. It switches calls (and sometimes data and messages) internally between telephones and also switches calls to the PSTN (Public Switched Telephone Network). PBXs save money by letting a large group of people (from about fifty to tens of thousands) share a smaller number of trunks, circuits or lines.

PBXs are modular or scaleable. To add more lines or stations, you add more cards to the cabinet (PBX box) and/or add more cabinets. Features are added via a software upgrade or by using an add-on PC that works as a server for the applications (like voicemail or IVR [Interactive Voice Response]) you're adding.

Each cabinet has a backplane with slots into which expansion cards are inserted. Communication takes place over the backplane, which sends signals from the cards to and from the lines and extensions. PBXs use the PCM (Pulse Code Modulation) protocol to turn analog signals from CO lines into frequency signals that are sent in binary code over the backplane of the PBX.

Connections between a PBX and its full-featured executive station sets are commonly digital, necessitating the purchase of compatible phones from the dealer you're using.

See the Secondary Market chapter for how to get manufacturer's phones for 70% off dealer prices.

Digital communications to the telephone station gives you improved signaling and phone control (fancy LCD displays, for instance), reduces ring-voltage requirements, and gives you dial tone and signaling over just one pair of wires.

Analog connections to PBX are used for regular single line telephones, modems, fax machines and adjunct systems that require them, like voicemail and IVR.

On the PBX's telco side, connections to the public telephone network are made through a variety of different cards and ports. Loop-start and ground-start analog trunk boards let you hook up standard analog Central Office (CO) trunks. ISDN BRI boards let you hook up ISDN Basic Rate circuits. T-1/PRI boards let you bring in 24-channel digital service.

The size of your company will determine whether to buy a key system or a PBX. The average life of a PBX is five to seven years. If your business will require up to 50 telephone stations over the system's life cycle, look at key systems. Over 50 stations usually justifies a PBX.

Digital Switching

Digital switching does not provide a continuous connection between the parties. While a call passes though a digital switch, the two lines are linked for brief flashes, thousands of times a second.

This process is called time division multiplexing (TDM).

Instead of transmitting the continuously varying signal of speech, digital systems sample the conversation (or other transmitted sound) 8000 times a second. The sound at the instant of sampling is then transmitted digitally as a binary number in a series of on-or-off pulses.

Speech samples are transmitted one after the other - but the system still returns to the first call within 1/8000th of a second. Typical digital switches can multiplex 30 or more conversations using TDM.

Once a sample has been read and passed to an outgoing speech storage location, another line is sampled. When 512 samples have been transmitted, the processor returns to the first line.

For more about TDM, see the T-1 Carriers chapter.

Anatomy of a PBX

A PBX can contain hundreds of computer cards, all of them designed to perform specific functions. The cards are inserted into slots on the system's backplane. The basic cards used by a PBX are:

Input/Output Cards (I/O Cards)
I/O Cards allow programming input and they send SMDR output to a printer or call accounting system.

Line Filter Unit
A Line Filter Unit takes as much background noise off of a circuit as possible. It is used when telco lines are brought in and when station line card signals are sent in and out of the PBX. Key systems do not usually include line filter units.

T-1 Interface Cards
T-1 interfaces in the PBX are configured for trunks, data, voice,

long distance company, 800, etc. but also must be configured by
the carrier so that what the carrier is sending matches what the
PBX expects to receive.

Tone Sender Unit
The tone sender unit passes touch tone signals along to circuits on
the interface cards or station cards. There are fewer TSUs than
station or line ports.

Tone Receiver Unit
Tone receiver units receives and interprets touch tone data
received from outside circuits and trunks.

Ring Generator
Ring generators send ring voltage (20 Hz, 90 Volts) down a telephone circuit to indicate that a complete circuit is being attempt-
ed. The telephone at the other end receives this ring voltage and
makes the telephone ring. When the call is answered, the circuit
is complete.

Manual switching systems always had an operator advising the
caller about the status of the call. Once the operator was removed,
another system was required to indicate call progress to the caller.

Progress tones are produced by the ring generator and include dial
tone, busy tone and reorder (or fast busy) tone.

Station Cards
Station cards can be analog or digital. They typically have eight or
12 telephone station ports each. Every digital or analog telephone,
fax jack or modem needs a station port in the PBX.

Trunk Interface Cards
Analog or digital interface cards accept incoming, outgoing or
bothway trunks from the telephone company. Each card has multiple ports for trunks, usually eight or 16 each.

PBX Features to Look For

PBX Interconnection

Two or more switches (and voicemail systems) that can be networked together via a point-to-point T-1 line. They can transparently transfer calls and messages to each other. When someone calls one site, but needs to speak to someone at the other office, they can be easily transferred by a receptionist.

Remote Call Forwarding

Remote call forwarding lets you forward an internal PBX extension to an outside phone number. Business calls can reach mobile workers and telecommuters, wherever they are.

Telephone systems that support this feature should allow the user to change the call forwarding settings from the systems' voicemail menu so that they can use the feature outside the office. If the settings have to be programmed from the office phone itself, it largely defeats the purpose of having the feature.

Remote call forwarding makes sense if you can consistently be reached at a single number, like your cell phone. If you are not at a predictable location, then "follow me roaming" may be a better choice.

You should have voicemail or an answering machine on the line you are forwarding calls to, otherwise forwarded calls may go unanswered. Most systems that support remote call forwarding can't take a message if the forwarded call is not answered at the remote number. The call can't be transferred back to the PBX or to its voice mail system once it goes off-premises.

Follow Me Roaming

Follow me roaming is like remote call forwarding, only the phone system can accommodate a list (usually four) of numbers that are dialed when the original extension doesn't get answered. The caller is put on hold while the system dials through the list.

There are two basic kinds follow-me dialing. With linear dialing, each number is dialed one after the other. With cascading dialing, all the "reach" numbers are called at once.

The system puts your caller on hold while it tries to reach you. If the call is not accepted or not answered, the caller goes to voice mail.

If the call is answered the system prompts for a code to accept the call before the call is transferred. This makes sure that only the intended parties get connected, not just anybody who picks up the follow-me call.

Make sure your PBX lets callers opt out to voicemail. Sometimes people just want to leave voicemail, not automatically go bouncing off to other numbers. As callers are waiting for follow-me connections, they are given the option to transfer to voicemail if they get impatient.

It's helpful if the PBX uses Caller ID or a password for follow-me screening. This way you can make sure that you get followed only by those callers you want to find you.

Make sure that callers can "follow" to internal PBX extensions as well as outside numbers-not all systems support this integration easily.

You should be able to schedule follow-me calls based on the time of day and the day of the week.

Off Premises Extensions (OPX)
Off premises extensions allow remote phones to work as if they were directly connected to the PBX in the office. Telecommuters

can get office-class conference calling, four-digit interoffice dialing, voicemail, fax and ACD (Automatic Call Distribution).

The goal is to make every phone in the company act as if it were connected locally to the main PBX. This means emulating a phone in front of the PBX and emulating a PBX at the phone. Often OPX-enabling PBX add-ons connect to the main switch via T-1 trunks and all call processing is handled by the PBX.

Computer Telephony Integration (CT or CTI)

Computer Telephony Integration lets computers talk to telephone systems so that callers can access databases, or desktop PCs can use Unified Messaging applications to get all of their communications on-screen.

> **See the Computer Telephony Integration chapter for much more about these and other CTI technologies.**

Call Announcing and Screening

Call announcing can be used for follow-me screening-the system asks callers their names, then broadcasts them to the intended parties when (and if) it finds them. They can then accept the follow-me connection or send the callers to voicemail.

Compatibility

PBX call progress tones need to be understood by third-party voice messaging and auto-attendant products. If the phone system is set for one signaling protocol and the adjunct system for another, the calls will be mishandled. Check to make sure that your existing add-on systems will understand, and be understood, by the PBX you're purchasing.

Toll Fraud Features

Find out what built-in measures the switch has to prevent unauthorized people from breaking into your PBX and dialing out over your trunks.

Remote Monitoring and Maintenance

Most PBXs support remote maintenance, so that your software, routing tables and area code tables can be updated without a technician sitting next to the machine. Remote monitoring lets your service company keep tabs on the system for you. With remote monitoring not only are you more likely to know sooner when your system has a problem, but the knowledge is come by relatively inexpensively.

Direct Inward System Access (DISA)

DISA is a feature that, while extremely useful, is the most often used vehicle for phone system fraud. DISA lets external callers dial into your phone system, enter a password, then be switched back out of the system, over company-paid long distance services.

How To Buy a PBX

Buying and installing a telecommunications system is complicated, but doesn't have to be difficult. The more information you have about the way your business operates, the better off you are. Likewise, the more you know about your options, the better your decision-making will be.

There are three sides to a telephone system, the wiring, the equipment and the service.

Look at your communications needs and how a new PBX and its associated applications can fulfill your requirements. Tell vendors how your business operates and how your employees use their phones.

First, decide what communications functions need to be performed in your organization. Once you know what you want to do, then look for the equipment that does everything you need to do for the best price.

By now you will have decided what features you need, what features you want, and you should know from this book what's involved in implementing them.

Ideally, unless you have an immediate need to cutover, you should give everyone possible an opportunity to give some feedback about the phones (and computers, if you're going CT) before you buy a system.

Before making a purchase, you need to properly understand your communications requirements. Present a list of your needs to vendors, and find out how they plan on meeting them. Follow this guideline to come up with an RFP to present to potential vendors.

Add-on technologies, like IVR and ACD, can justify, and sometimes pay for, the cost of a new PBX. IVR automates routine calls, freeing up employees to do other, more important work. ACD efficiently distributes calls, making an office or call center more productive. Both technologies can also reduce talk time, saving you money on toll-free lines and toll charges. Efficient communications in today's business world dramatically affects the bottom line-how much money your company will make.

Plan to spend anywhere from $200 per user to $1000 per user depending on the type of system you are interested in.

PBX Present and Future

The PBX is known for reliability, has been around for quite some time and will still be here in years to come. To get the most out of a PBX system, know what you need, know what your proposed system can do, then apply those system features to the operation of your business to get the most out of a PBX and the business.

Three technology trends are leading to momentous changes PBX architecture and design

"Open" PBXs and IP Telephony Servers Proprietary links from PBXs for LAN-based applications are becoming a thing of the past as the PBX industry moves to adopt the Internet Protocol. Open systems that support IP will allow direct support of LAN-based applications and CT (computer telephony) servers.

ATM Many PBX suppliers have jumped on the ATM bandwagon, even though customer demand is lagging. ATM networks are still used primarily for larger integrated voice/data networks.

IP Transmission and Routing While TDM/PCM circuit-switching remains the most reliable and consistent platform for voice, there is much work going on to develop Ethernet-based, packet-switched voice. Ethernet-based PBXs are about to hit the market.

The past five years have seen dramatic improvements in PBX system performance, both in traffic volume and call processing capabilities. PBX processing will get even faster when ATM switching is commonplace.

Communication Servers

Communications servers (also called PC PBX or CTserver or UnPBX) are more than PC-based alternatives to key systems and small PBXs.

Comms servers provide lots of call handling in one box. In addition to basic dial tone, comms servers can integrate PBX-style features, computer telephony functions, and run everything alongside data over a PC network.

You can get (internal and external) email, voicemail and fax messages on your computer screen, called "unified messaging." You can prioritize your work much more effectively when all of your messages are in one place and can be handled together.

Some comms servers also include auto attendant/voicemail, IVR (Interactive Voice Response), ACD (Automatic Call Distribution), unified messaging and IP telephony applications.

Most communications servers are PC-based, but some, made by phone system vendors, are actually phone systems modified to act like PCs.

A comms server works well for communications-intensive businesses (like brokerage firms, high tech startups and retail stores) and companies with busy data networks and/or a large number of branch offices. If the comms servers are in the branch locations, you can send voice calls over the same network you send data, and save money on long distance phone calls.

Another good place for a comms server is in a corporate department, for ACD (Automatic Call Distribution). Departmental ACD applications can be expensive to implement if the corporate PBX has to be upgraded. A local comms server for ACD can be a real bargain.

Capacity limitations prevent large companies from using comms servers as their primary telephone system. Upper port limits are currently at about 500 combined voice, data and trunk ports.

Comms servers have boards that accept phone lines from the outside public switched telephone network, or PSTN. Network interface cards can be analog or digital, and range in size from two ports to eight ports. There are separate expansion cards that handle ports for telephones. Multi-service boards can provide speech recognition, conferencing and fax on a single board.

PC-based comms servers are expanded by connecting PCs together usually using TDM (Time Division Multiplexing) across the PCs' communications buses.

The biggest advantage of a PC-based phone system is flexibility. A PC-savvy user can program and maintain his or her own phone system and have more control over how calls are handled. Comms servers systems typically ship with menu driven or GUI (Graphical User Interface) software which make their use relatively straightforward for somebody with PC troubleshooting and administration experience.

Types of Communications Servers

LAN based Servers

These servers route calls from one to another in a distributed server network across a company LAN (Local Area Network) or WAN (Wide Area Network). Collectively they provide switching, call control, interactive voice response and messaging functions. The failure of an individual server affects only those users connected to it.

One of the greatest benefits of distributed telephone systems is fault tolerance. Because each server on the network can operate independently of the others, the distributed phone system will have no central point of failure. A telephone network running on top of a switched Ethernet network (which has multiple paths) is more redundant than a centralized switch, where the failure of the single processor can disrupt all the users.

LAN based servers can support several processing functions. One server can concurrently run switching, IVR, voice mail and email services. If you can't (technically or economically) work with servers, then a network appliance is probably better for your business.

Network Appliance

The appliance works like a data hub or router. Network appliances create a local area network (LAN) for connecting PCs, printers, servers and other network devices, and are designed for small and medium-sized businesses or branch offices.

Network appliances typically include built-in voice mail, a network call processor for call control, auto attendant, browser administration, TAPI for connectivity to standard telephone systems and the capability to hook up to the public switched telephone network (PSTN) plus wide area network (WAN)/Internet connectivity for intra-office communications.

The network appliance box has a bank of RJ14 modular analog jacks that connect to desktop telephone handsets, and RJ45 modular jacks that connect to the company LAN/WAN via Ethernet or ATM.

Network appliances are less likely to fail than a PC because they are solid state devices with no moving parts and generally have simpler operating systems.

Network appliance/router-based systems sell for $550 to $600 per user, more if you use speakerphones or multi-line phones at the desktop.

IP-based PBX

An Internet Protocol PBX is different from a PC-based comms server in that it has IP telephony gateways instead of trunk and line cards to interface with the world outside your office.

With an IP based PBX your voice calls are sent using the same packet-based protocol that the Internet uses. Voice over the Internet replaces traditional dial tone and Central Office switched services, and can be cost effective between distant offices (and other continents) because telephone calls are free between the IP gateways. Expansion is accomplished through the TCP/IP network; the more traffic you have, the more bandwidth you use. Remote office phones and PCs plug into Ethernet ports with local IP gateways. Calls can be routed so that they enter the PSTN where the call will be least expensive to complete.

> See the *Voice Over the Interntet Protocol (VoIP)* chapter for more about IP telephony and voice calls using the Internet Protocol.

Peer to Peer Networks

These networks require companies to replace all of their phones with H.323 Ethernet phones. H.323 (this defines a standard for IP telephony) and SCTP (this defines a standard interface for call control and station management services) make full-featured digital desk phones possible. H.323 phones are available now, but are very, very new.

H.323 requires separate PC servers for enhanced telephony services like voice mail, IVR, and ACD, however, there is no dependency on PCs for basic telephony services because the H.323 network doesn't require (or use) traditional dial tone and telephone signaling. The peer to peer networking approach using H.323 will work for smaller, geographically distributed companies that are connected over a WAN.

Comms Servers Today and Tomorrow

If all you want is a key system for your ten-person office and want to spend under $2,000, this is not the technology for you. But if you're looking to beef up a phone system with adjunct systems, a PC based telephony server could be cheaper than a traditional phone system plus the adjuncts.

The ideal comms server, with PBX call handling and features, an IP (Internet Protocol) telephony gateway, voice mail, IVR, speech recognition, and other advanced applications, is still not a mainstream product. Most vendors have so far focused on specific applications or features.

CT vendors admit that in reality a full-featured comms server will probably be a network of servers, each hosting a separate application, and they also always recommend installing extra servers for backup. PCs are still not as reliable as traditional telephone systems.

Distribution is the biggest barrier to the success of comms servers in the telephone system market. There are basic voice telephone system distributors, and different types of data distributors. The market hasn't yet decided who's going to sell PC-based phone systems.

Another problem with PC PBXs is that they aren't yet out-of-box products. They are generally built to order, and their price reflects it.

Comms servers are limited in expansion capabilities to the size of the PC box, the number of PCs that can be chained together, and to the number of ports that can be crammed onto a card. Today's maximum capacity of about 500 ports is changing fast, though, as higher-density boards are developed.

Recent advances mean that comms servers can handle most Central Office trunks, and peer-to-peer applications have real value for companies that are data-intensive and/or have offices in several locations. If you decide to buy a communications server, pay close attention to the relative costs of digital vs analog components and services. A change in the technology you use could cost big bucks without giving you a comparable increase in utility.

Repeaters and Routers

Repeaters

Repeaters extend transmission of a signal over a distance. They're used to interconnect segments or paths within in a local area network (LAN). They're also used to amplify and extend wide area network (WAN) transmissions.

A repeater receives a signal over a copper or optical cable, amplifies that signal, and then retransmits it.

In addition to strengthening the signal, repeaters also remove the "noise" or unwanted aspects of the signal. Repeaters fix signal attenuation problems caused by external noise or cable loss.

Digital signals depend on the presence or absence of voltage and suffer from attenuation more quickly than analog signals. Digital signal repeaters are typically placed at 2,000 to 6,000 meter intervals, analog at 18,000 meter intervals.

If the physical medium is copper cable, the repeater is an amplifier circuit and two signal transformers. The impedance of the cable must be matched to the input and output of the amplifier. Impedance matching also minimizes signal reflection along the cable, thereby reducing echo.

Fiber optic repeaters combine a photocell, an amplifier, and a light-emitting diode (LED) or infrared-emitting diode (IRED) for each signal that requires amplification. Fiber optic repeaters require careful design to ensure that internal circuit noise is minimized.

Routers

A router is a smart switching device that receives packet-based data, determines the next network node to which the packet should be sent, and transmits it to its next destination. A packet may travel through several routers before arriving at its destination.

Routers create or maintain a table of available transmission routes and their conditions. This information, along with distance and cost algorithms, is used to determine the best route for a given packet.

Located at network junctures or gateways, including at Internet points-of-presence (POP), routers are often integrated into network data switches.

Routers can be used for advanced communications including:

Switched Virtual LANs
Routing is shifting from routing between physical parts of a LAN to routing between software-defined logical virtual LANs.

Networked Multimedia
Routing multiple media types requires a high quality of service and high-performance advanced queuing techniques that routers can provide.

Internet Access
The explosive growth of the world's Internet continues to drive high-performance routing requirements.

How Routers Work

Routers contain network interface modules that connect the router to local (LAN) and wide area (WAN) physical networks. They also have shared memory (used to store packets during processing), a CPU, and storage for configuration files, switching caches, routing tables and the manufacturer's operating system.

The first thing a router does when it receives a packet from the outside network is discard the Layer 2 (data link layer) frame information in which the Layer 3 (network layer) packet is encapsulated and place the part of the packet that remains into its memory.

> See the *Data Communications, Digital Voice and the OSI Model* chapter for an easy-to-understand, complete description of the OSI model. The OSI model organizes data communications functions in terms of layers.

The CPU checks the switching cache for the information it needs to make a network layer switching decision about the packet. If the information isn't there, the router drops into process-switching mode, and the CPU gets network switching information from the routing table in its memory.

When it gets the switching information it needs, the router encapsulates the packet in a new data link layer frame, updates any packet counters as necessary, and then sends the packet on its way through the outbound interface.

After it reads the first one, the router fast-switches all subsequent packets with the same destination address. When it's fast switching, a router stops whatever else it's doing and forwards the packet immediately, making fast switching speedier and more efficient than process-switching.

When a router drops into process-switching mode, it returns to its scheduled task list and doesn't forward the packet in question until packet forwarding comes up again on its list of tasks.

Some protocols need to be process switched, including most IBM protocols and every X.25 transmission. If you're running these protocols, you should consider using switches that support high process-switching speeds.

There are three ways to optimize your router's performance: buy fast-switching routers, use your routers as efficiently as possible, and optimize your bandwidth between your routers.

Bandwidth can be dynamically manipulated to maximize data rates and minimize congestion. Load balancing can be used with parallel physical circuits or combined ISDN or T-1 channels. You can change on the fly how much of the circuit's bandwidth you use for a particular transmission.

Consider the following when buying a router:

Performance Requirements
Understand your company's requirements. Make sure you have enough processing power available for packet switching and memory handling.

Network Access and Switching
Use the fastest supported switching path you can afford.

Special Features
Identify your needs for features such as compression and encryption. Implement them carefully, so they have limited impact on switching performance.

Centrex

Centrex (also called Plexar, CentraNet, Centron, Cenpac, Intellipath or Presstige) is a business telephone service offered by your local telephone company from the local central office. Most medium-sized and larger companies use a PBX because it's much less expensive than connecting an external telephone line to every telephone in the organization. In addition, it's easier to call someone within a PBX because the number you need to dial is typically just 3 or 4 digits.

But many companies, particularly smaller businesses, may not want to purchase and manage their own telephone system because of the capital investment, technical requirements, or time limitations. Most local telephone companies sell Centrex and lease it to businesses as a substitute for an on-the-business-premisess telephone system that must be bought or leased.

Centrex lines usually cost about 20% to 50% more per month than plain analog phone lines. Since local dial tone in the US and Canada is inexpensive, Centrex can be a cost effective way to get the features of a PBX without having to buy a PBX.

Centrex may not be for you. You need to compare the rates and features of Centrex to what PBX and key systems have to offer.

You need to decide what makes better business sense: leasing Centrex or buying your own switching system.

The principal advantages Centrex to small and home-based businesses (called SOHO or Small Office, Home Office) are a low cost of entry, centralized maintenance, and scalability.

You get "business phone service" without having to buy equipment (though there are contracts and recurring charges). You can tie together multiple locations under the same four-digit dialing scheme (though many PBXs can also be configured to do this). You don't need to hire somebody to babysit your phone system. And you can (if your CO has the capacity) expand from five lines to five thousand, without running through several generations of CPE (Customer Premisess Equipment) in the process.

The Benefits of Centrex

The Centrex telecommunications system is owned, maintained and housed by the phone company. They phone company is responsible.

Your telephony services grow with you and you can customize them to match your business needs.

Centrex is kept state-of-the-art; software and hardware upgrades are ongoing at no cost to the consumer (although additional features cost more per month).

Multiple locations within the same CO (Central Office) can be supported without buying OPX (Off Premisess Extension) services or equipment.

Telephone company salespeople can be very persuasive, so think carefully before buying Centrex. One-person offices can benefit from its features, but several lines generally require fancier and more expensive equipment than a single line set to effectively handle calls. Once you move from basic sets, the cost advantage of Centrex is generally lost.

Basic Centrex Features

Call Transfer transfers calls to another line either inside or outside your Centrex system.

Direct Inward and Outward Dialing allows you to control whether incoming or outgoing calls are routed directly to an employee's desk or to an attendant.

Hunting automatically sends an incoming call from a busy line to the next designated line.

Line Restriction limits phone access on selected lines, so that only authorized numbers or regions can be called.

Station Line Identification provides a record of calls made by each extension.

Station-to-Station Dialing allows station users to intercom between stations by using abbreviated dialing, usually four digits.

Three-Way Calling allows you to add a third person to a regular two-way call.

Optional Centrex Features
(Available at an additional cost per line)

Call Forwarding-Busy Line automatically routes calls to another extension or to your voice mail if your line is busy.

Call Forwarding-Don't Answer automatically reroutes calls to another phone if you haven't responded within a preset number of rings.

Call Forwarding-Variable sends your calls immediately to another number and is set by the user by entering a three digit code and the destination number.

Call Hold puts a call on hold and returns dial tone.

Call Park lets you hold a call on one Centrex line and pick up a call on another.

Call Pickup lets you use your extension to answer any ringing phone in your designated group.

Call Return lets you return your last incoming call with a code or a single button.

Call Screen allows up to 10 predetermined numbers to be routed directly to a prerecorded announcement.

Call Waiting-Incoming signals you with a tone when another call is coming through on your line.

Caller ID displays the caller's phone number (on compatible equipment).

Directed Call Pickup use your extension to answer a specific ringing phone in your designated group.

Direct Inward Dial to Direct Outward Dial Transfer allows you to transfer an incoming direct-dialed call to a location, either within, or outside the Centrex system. The caller doesn't have to hang up and dial the new number.

Distinctive Ringing lets you know if a call is originating inside your office. A single ring means an inside call, a double ring means the call came from outside.

Executive Busy Override permits an interruption on a station that has a call in progress. Parties engaged in conversation hear a warning tone before new caller joins the conversation.

Flexible Route Selection automatically routes calls over designated carriers. You can cut costs if your business does a high volume of statewide or nationwide calling by using your preselected carrier.

Meet Me Conference allows you to meet with a maximum of either 6 or 30 people on a conference call.

Message Waiting Indicator alerts you with a stutter dial tone when someone has left a message for you in your voicemail box.

Music On Hold provides music or an announcement to callers on hold.

Paging provides access to customer-owned paging systems.

Priority Ringing special ringing indicates when any of up to ten numbers you select are calling. If you also subscribe to Call Waiting, a unique call waiting tone alerts you that the caller is one you've selected on your Priority Ringing list.

Remote Access to Call Forwarding allows you to control remotely the destination of your forwarded calls.

With **Repeat Dialing** you can program your phone to keep dialing when your call can't go through. A special ring lets you know when the line is free.

Sectional Billing permits you to group individual Centrex lines by department, agency or work group.

Select Call Acceptance only callers whose numbers are on your selected list will be allowed to complete the call. Automatically accepts calls from up to 31 telephone numbers you choose.

Select Call Forwarding forwards calls from up to ten numbers you select to any number you designate. Important calls can reach you directly, rather than forwarding to voice mail. Calls not on your special list can be forwarded to yet another number by using Call Forwarding/Variable, Call Forwarding/Busy Line, or Call Forwarding/Don't Answer.

Speed Calling (also called abbreviated dialing) lets you press fewer digits to make a call.

Centrex Buying Tips

Decide how you want to set up your phones. Do you want to run the system like a PBX (where calls come in at a central number and are transferred to extensions by a human or mechanical attendant) or make extensive use of Centrex' DID (Direct Inward Dial) capabilities to give everybody a number? If you go with a PBX-type setup and use a receptionist, you'll probably need an attendant console.

If you go with a departmentalized setup (clustered workgroups like sales, accounting, administration, etc.) a KSU-less Centrex-compatible phone system can give you a non-squared configuration. This provides private lines on multi-line phones, common lines and intercom groups on a department basis.

Look for Centrex CPE (Customer Premisess Equipment) that gives you some control over Centrex features. Some of today's CPE lets you control music-on-hold selections, call accounting, overhead paging systems, and station displays.

Smaller sites (roughly five to 75 stations) can usually get package deals on Centrex services from RBOCs (Regional Bell Operating Companies). Ask about these deals, that include services like conferencing, transfer, call forward and more, if you're considering Centrex for a relatively small number of stations.

Look for Centrex CPE (particularly attendant positions and desk phones) that also supports digital lines, like ISDN. You will have a wider range of applications to choose from.

Many RBOCs are offering ISDN as an enhancement for Centrex. Use digital ISDN lines for fast, high-bandwidth intensive applications (data and videoconferencing) and the analog Centrex lines for traditional call processing, voicemail, etc.

Other recent Centrex enhancements include wireless capabilities, automatic call distribution, voicemail, fax services, call accounting and PC-based call routing systems.

When your switch is inside the CO (as it is with Centrex), it's harder to provide yourself with modern phone system adjuncts like unified messaging, operator consoles, call accounting and other facilities.

This is where Centrex CPE steps in. There is plenty of Centrex CPE available, and the products are starting to support advanced applications and equipment.

Types of Centrex CPE

Centrex Phones
Centrex phones let you hold, transfer and forward calls by simply pressing a button. Many have message waiting lights that flash for stutter dial tone (CO voicemail) and offer other "smart" features.

Call Accounting.
Centrex-compatible call accounting packages bring Message Detail Records straight from the CO to your desktop. You can use this info for charts, graphs and reports. These packages monitor the CO

117

directly via a CO-supplied SMDI (Station Message Detail Interface). Alternatively, there are site-monitoring boxes that live in your office, monitor your trunks and deliver SMDR records via a serial interface, just like you get from a conventional phone system.

Voicemail

Central Office voicemail is fine for small businesses, but is not flexible enough for larger firms requiring auto attendant functionality, dial-by-name and other applications. Fortunately, most general-purpose voicemail systems can be used with Centrex so you have lots of choices.

ACDs

Centrex-compatible ACDs let you queue calls, manage agents and generate realtime reports-all from CO trunks.

Paging Systems

Centrex-compatible voice paging systems give you all the features of classic, switch-based paging equipment.

Attendant Consoles

Centrex answering consoles can take many shapes; multi-button phone, conventional-appearing attendant console, or PC-based system.

Set Handlers

These systems convert analog or digital Centrex service to proprietary digital formats and distribute the service to full-featured digital station-sets. They employ CO features to handle transfer, forward and other functions, while giving you much of the appearance and functionality of an onsite digital switch.

Administration Systems

Centrex administration systems let you reconfigure your Centrex service. Platform-based consoles let you turn enhanced CO services (such as CO voicemail) on and off; activate or deactivate lines; add or reassign trunks and rearrange long-distance dialing patterns to route calls over the least expensive carriers.

Multiple Site Centrex

Almost all of the above categories of CPE come in both single- and multi-site versions. If your company has multiple locations, you should get the appropriate CPE.

Specifying Centrex Services

You'll need this information to be able to get Centrex pricing or to place a Centrex order.

Telephone Service For Individuals (Employee Extensions)

How many analog extensions (lines) do you need in your office?
Do you want voice mail from the phone company? If so, on how many of these extensions?
Do you want Caller ID?
Do you want three way calling on these lines?
Do you want call transfer capability on these lines?
Other features desired
Special requests or comments

Primary Voice Numbers (Main Telephone Number, Receptionist Switchboard, Auto Attendant)

Here you are requesting phone lines to ring into your receptionist or to an automated attendant system that automatically transfers calls to personal extensions.

How many incoming lines do you need?
Will you be using an automated attendant to answer incoming calls?
Do you want to request a vanity number for your business?
Do you want the telephone company to answer with their voice mail service if all incoming lines are busy or unanswered?

Analog Data/Fax Lines (Fax Machines, Analog Modems, Credit Card Terminals)

How many fax lines do you need? (If more than one, do you want them to be put in a hunt group?
Do you want Caller ID on your fax lines?
How many lines do you need for modems?
Do you want Caller ID on your data lines?

Living With Centrex

Once you decide to use Centrex service, be very meticulous about the ordering process. Take notes when you're on the phone with the phone company - write down the date and time of the call, what was discussed and the name of the person you talked to. Ask the representative the name of the backup person when your rep is not available. Verify your schedule regularly, right up to system installation.

Call Center Technology

Call Centers, Automatic Call Distribution (ACD), and Predictive Dialers

A call center is a group of people, called agents, who, with their associated phone and computer equipment, do the same, repetitive, type of work.

Inbound call centers handle incoming calls (and faxes, email, and queries from Web sites), and outbound call centers make outgoing calls, usually telemarketing or research-related.

Inbound Call Centers and ACD</B>

Most traditional call centers handle inbound voice calls. Special software and software/hardware combinations for call centers called Automatic Call Distributors (ACDs) distribute these calls to agents that can handle them using any of a number of call allocation and distribution schemes.

An ACD does two things. It minimizes the amount of time that a customer spends waiting to be helped, and it maximizes the amount of time that agents spend interacting with customers.

ACD systems are becoming more sophisticated and are competitively priced. Even small businesses can enjoy the benefits of logical distribution and management of inbound calls.

Once a call arrives at an ACD, sometimes directly from the outside telephone network, sometimes from the internal phone system or IVR (Interactive Voice Response) system, the ACD distributes the call to whoever is available in the group that is handling those particular calls.

The ACD system decides which phone will ring next and puts overflow calls into a holding pattern until someone is available. This holding pattern (or queue) keeps track of the order of the callers, plays music-on-hold and can play update greetings for the callers. Update greetings assure the caller that he has not been forgotten or can give him the option to leave a voicemail message instead of continuing to wait.

Some ACD systems can inform the caller of the average wait time in the queue or update them whenever their position in the queue changes. This feature lets the callers to judge whether or not they want to continue holding and puts them more in control of the situation.

A small office can use an ACD system to evenly distribute calls from prospective clients to sales representatives or to balance the incoming call load between all employees (or a designated group) in the office to ensure that all calls get answered immediately.

You could configure your ACD system to distribute all incoming calls to all employees, for instance, only if the receptionist (if you have one) is on another call.

Some systems can send the customer's data file with the voice call, so the file reaches the agent's computer as the call reaches her headset. Some ACD packages let callers punch in a phone number at which they wish to be called back, hold their place in the queue, then automatically call them back when an agent is available.

An ACD integrated with the Internet displays a button on the associated Web page. When a caller clicks on the icon, they're usually given a screen-based form to fill out. The customer can request an immediate callback from an agent or a callback at a later time.

Internet applications with video capability are available on ACD systems, usually larger systems, and they require the customer to have a video camera attached to his computer. Video technology that provides fairly respectable smoothness of motion is possible at 128 Kbps (two channels of a BRI ISDN line). Internet-enabled agents can push Web pages to callers and/or use text chat.

Agent Callback lets an agent flag calls, and the system uses ANI (Automatic Number Identification) so if a caller dials back within a set period the same agent can receive the call.

Routing Models

Which call goes to which agent? Choosing the right routing strategy is crucial to obtaining high quality of service and managing staff.

The simplest kind of routing sends the oldest call in queue to the next available agent; a strategy that works fine if traffic is very uniform and agents are equally efficient in handling calls. Otherwise, this "next available agent" (NAA) queuing will overburden your most efficient people.

An improvement on NAA entails sending the oldest call in queue to the longest-idle agent, a slightly more intelligent strategy that helps balance call loads. Basic ACDs support both strategies.

The next step up lets you establish groups of agents with a common purpose and similar skills. You can send calls to each group separately, based on choices the caller makes. Inbound calls are sorted through an auto attendant or IVR (Interactive Voice Response) system.

An improvement on this is called fallback routing. Calls awaiting a particular group become accessible to a more general pool of agents, so as time passes, callers don't have to wait if their particular agent pool is busier than others are.

The most intelligent routing is "skills-based" where multiple criteria play a role in matching callers to qualified agents.

Here's how it works: a call comes in and finds that all agents are busy. Instead of getting to a traditional first-in-first-out queue the call waits in a pool of calls.

When an agent become available, the ACD looks at the pool of calls and decides which one should be assigned to that particular agent. The assignment can be made based on the importance of the caller to your business, the caller's willingness to wait, a skills match between the agent and what the caller needs, or a combination of these and other criteria.

Callers are identified by their telephone numbers, the number they dialed, or by a pin number they enter into an IVR front end.

Managing an installation setup around skills-based routing is tough because you have to worry about staffing. How do you schedule agents who are "experts" rather than interchangeable?

Types of ACDs

ACDs can be a part of your PBX software, based in a separate PC (called a switchless or software-based ACD) and attached to a phone system, or stand-alone, meaning a switching capability is built into the system and it processes calls by itself.

Standalone ACD systems are sometimes known as ACD/PBXs, because they combine ACD call handling with the functions of a PBX. If your company is primarily a sales office or you plan to make money providing call center services, this kind of system might be better for you than a PBX by itself. Adding ACD software to an existing key system or PBX later might be more expensive than buying an ACD/PBX combination initially.

Virtual ACDs give you ACD features, but require no customer premises equipment. Your incoming calls are processed outside your facility and you pay for usage only, typically between 20 to 25 cents per minute.

PC-or server-based systems are aimed at under-400 person call centers that have significant data requirements. Comms server-based ACDs are easier to interface to information databases because they're PC-based and can communicate digitally with other computers.

> **For more information about communications servers, see the *Communications Servers* chapter of this book.**

Scaling Issues

A conventional phone system (PBX) lets a large group of people share a smaller number of CO lines. The assumption is that everybody isn't going to be talking on the phone at the same time.

An ACD - a call center phone system - makes the opposite assumption. You want everybody to be talking on the phone all the time, which means you'll need at least as many trunks as you

have agents. You'll also need to queue calls and keep them on hold until an agent becomes available. There are several ways to do this.

The simplest way is to rent more trunks than you have agents; and buy premises equipment that can terminate these trunks, queue the calls they provide, play clever on-hold messages to keep callers from hanging up, and offer automated facilities (voicemail, IVR, etc.) to let some callers help themselves.

This works fine when traffic to your center is relatively constant - buoyed up by frequent catalog mailings, regular advertising, or other publicity.

But heavy seasonal variation, promotional campaigns, the need to blend inbound and outbound calling, serve callers from multiple time zones, etc., can throw a wrench into your ability to predict traffic and provide facilities for dealing with it. If you anticipate this kind of situation, it makes sense to explore call center systems that leverage the public network to provide greater call handling flexibility.

At the simplest level, both digital Centrex and ISDN service let a single line manage multiple call appearances. When you set up a Centrex- or ISDN-based ACD system, inbound calls can be held at the CO until agents are available. Waiting calls can be polled, played messages and music, and given the chance to transfer to IVR, voicemail, or other facilities - all without you shelling out for extra hardware, physical ports, and trunks to your premises.

At a more global level, you can use specialized CPE (Customer Premises Equipment) and network services to create a "virtual ACD" that distributes calls among multiple physical call centers according to different criteria: availability, time of day, the call's point of origin, or other variables. The most obvious benefit of this approach is that it lets you tie remote agents together for a seamless-appearing center, while saving money on overhead.

Given the choice between waiting in a hold queue and writing email, people often opt for the latter. But those "info@" or "support@" addresses on a Web site are seldom monitored efficiently.

Over the past year or so, however, a number of companies have created a major breakthrough. They're building software that routes inbound email around a call center and queues, prioritizes the traffic, lets agents respond to emailed queries using semi-boilerplate messaging, and keeps stats on the whole process.

Here are some things to know about inbound call centers:

One: Calls Get Bunched Up

Call come in groups, but fortunately, arrive in predictable repeating patterns by time of day, day of week and season, and can be easily forecast, (for call center staffing and other resource planning) down to specific half-hour increments.

But even with near perfect forecasts, the actual moment-to-moment arrival of calls within the half hour is a random phenomenon. Performance objectives and standards must take random call arrival into account. A "per day" type objective doesn't make sense when random calls mean unpredictable wait times for agents. (Unless the call center is badly designed and always backed up.)

Two: There's a Direct Link Between Resources and Results

Service level is defined as "X percent of calls answered in Y seconds: such as 90% answered in 20 seconds.

The level of service rises geometrically until the service level is obtained. Economists call this the "law of diminishing returns."

Each additional agent has a smaller and smaller incremental effect on the service level.

Dramatic improvement may not require a lot of resources. On the other hand, those who want to be the best of the best find that it takes a real commitment in the staffing budget. And, the call center budget should reflect the level of commitment.

The service level affects the quality of the service the agents provide. Customers complain, calls get longer, agents get dissatisfied and less friendly. There's a circle between service level and quality.

Three: When Service Level Improves, Productivity Declines

As service level goes up agents handle fewer calls each. Put another way, as service levels go up, occupancy goes down. Occupancy is the percentage of time during a half-hour that those agents who are on the phones are involved in talk time and post-call work. The inverse of occupancy is the time agents spend, plugged in and available, waiting for inbound calls.

If occupancy is high, it is because the agents on the phone are taking one call after another and another, with little or no wait between calls. When the service level gets better, occupancy goes down. Average calls per individual also go down.

This loss of productivity is counteracted in most call centers by having the agents do non-phone tasks when the inbound call load slows down. Called a "blended" environment. But understand that when non-phone work is getting done, either there are more agents assigned than the base line service level indicates, or the service level objective is sacrificed because calls aren't getting handled as well as they should be.

Four: You Need More Staff on Schedule Than on the Phones

You can miss your service objectives by a lot because you don't have the staff you expected in the right places at the right times. Many things that can keep agents from their desks; lunch, breaks, meetings, etc. A strict 15 or 20% over base recommended staffing levels won't work unless you figure in variations like the fact that agents are most often absent before and after weekends.

Five: Staffing and Telecom Budgets Must Be Integrated

Staffing and trunking issues are inextricably associated but are often paid for out of different corporate budgets. Staffing levels also affect expenses such as toll-free costs.

Six: The Demands On Call Center Agents Are Increasing

Repetitive information transmission is increasingly automated, and consumers are well educated, informed and equipped. There is a "skills gap."

Call centers are an important part of a much bigger company-wide process. Coordination with other departments and integration of the call center with other parts of the organization is important.

Buying an ACD

You don't have to tackle everything at once - or at all, if the scale and goals of your center don't warrant it. Let your center evolve as you gradually acquire and integrate call handling products and solutions.

This is more and more feasible, since today's ACD systems are largely software-based, often run on PCs, and will communicate with agent PCs, IVR, databases, and other outside resources over relatively simple and increasingly standardized interfaces.

Outbound Call Centers and Predictive Dialers

Predictive dialing is a technology used in call centers for outbound dialing, usually telemarketing. Predictive dialers are inexpensive and can be used by even the smallest companies.

Predictive dialers automatically outdial phone numbers from a pre-defined list. The calls are made (from a database or callout list) by the predictive dialing software and passed to a live agent only once a person has answered. This saves the agent from dealing with wrong numbers, answering machines, and calls that are not answered.

A predictive dialer can manage all the tasks associated with making a call. They can queue new calls by using statistical averages to predict when an agent will complete the current call. This technology enables the dialer to begin dialing a new call before the current call is finished.

Predictive dialers can further increase your business' efficiency by simultaneously delivering the voice call to a telephone (or headset), and that consumer's record to an agent's PC. This increases the average productive talk time per agent, per hour.

Phone System Equipment List

Manufacturer	System Type	System Name	Max Lines	Max Ports
AltiGen Communications	CommSrvr	AltiServ	192	192
Amtelco	CommSrvr	ES/BX Infinity	20	100
Applied Voice Technology (AVT)	CommSrvr	AgentXpressNT	48	84
Artisoft	CommSrvr	TeleVantage	n/a	96
COM2001 Technologies	CommSrvr	Network Telephony Xchange (NTX)	48	48
Computer Talk Technology	CommSrvr	Intelligent Call Exchange (ICE)	136	72
Harris Corporation	CommSrvr	IXP Communications Platform	4,608	4,608
Interactive Intelligence	CommSrvr	Enterprise Interaction Center	350	264
Inter-Tel	CommSrvr	Call Server	258	258
Mitel Corporation	CommSrvr	SX-2000 for Windows NT	96	176
NBX Corporation	CommSrvr	NBX 100	48	144
Nexus Telecom	CommSrvr	Call Server	24	144
TouchWave	CommSrvr	WebSwitch	8	16
OmniLink Communications	ISDN TA	NetPacer	4	20
Cortelco Systems	ISDNPBX	Millennium Digital Communications Platform	1,024	1,024
AVG-Eagle	Key System	Eaglet	6	16
AVG-Eagle	Key System	Eagle/One	32	64
AVG-Eagle	Key System	Digi-Eagle	64	96
BBS Telecom	Key System	Plexus	120	128
Comdial	Key System	Voyager	4	8
Comdial	Key System	Unisyn	6	16
Comdial	Key System	Impression	224	192
Cortelco Kellogg	Key System	Odyssey	24	120
Cortelco Kellogg	Key System	Aries Digital ADKS-144	96	144
Cortelco Kellogg	Key System	Aries Digital ADKS-308/616/924	9	47
Creative Integrated Systems	Key System	Teligent 314	3	14
ESI	Key System	IVX	16	34
Executone Information Systems	Key System	Integrated Digital System (IDS)	648	648
Flash Communications	Key System	Flash-Com System	2	16
Fujitsu Business Communication Sys	Key System	Allegra	96	192
Inter-Tel	Key System	Axxent	12	48
Iwatsu	Key System	ADIX-M	128	120
Iwatsu	Key System	ADIX-S	28	32

KS Telecom	Key System	Atlas ISDN 500	120	460
Lucent Technologies	Key System	Partner Advanced Communications System (ACS)	15	40
Mitel Corporation	Key System	SX-50	180	160
NEC America	Key System	Electra Professional 120	64	96
Nitsuko America	Key System	Portrait 308	3	8
Nitsuko America	Key System	Portrait 824	8	24
Nitsuko America	Key System	124i	52	72
Nitsuko America	Key System	DS01	24	72
Nortel	Key System	Venture	3	8
Nortel (Northern Telecom)	Key System	Norstar Compact ICS	16	24
Nortel (Northern Telecom)	Key System	Norstar Modular ICS	120	128
Panasonic	Key System	KX-TD1232/816/308	24	128
Panasonic Telecommunication Sys	Key System	DBS	32	72
Picazo Communications	Key System	VS1 Business Telephone System	96	192
Samsung	Key System	Prostar 816 Plus	8	16
Samsung Telecommunications	Key System	DCS Compact	10	32
Samsung Telecommunications	Key System	DCS 50si	48	40
Sprint Products Group	Key System	ProtÈgÈ	72	72
Tadiran Telecommunications	Key System	Coral SL	168	128
Teleco	Key System	UST DK 1000 Series	200	336
Telrad	Key System	Synopsis	8 .	25
Telrad Telecommunications	Key System	Digital Key SystemBx	144	254
Teltronics	Key System	VisionLS	288	288
Toshiba	Key System	Strata DK40	12	28
Toshiba	Key System	Strata DK14	4	10
TransTel Communications	Key System	SK-408	4	8
TransTel Communications	Key System	SK-824	8	24
TransTel Communications	Key System	Superkey SK-824	8	24
Vodavi Technology	Key System	STARPLUS Electronic Key System	40	96
Vodavi Technology	Key System	STARPLUS DHS	12	24
dba Telecom	KSUless	SmarTalk NRG	3	n/a
Newtronix	KSUless	Success Phones	4	16
Telematrix	KSUless	TMX	5	8
TMC	KSUless	SOHO	4	24
TT Systems	KSUless	SOHO Phones	4	16
Bosch Telecom, Inc.	PBX	Bosch Integral	16,000	32,000
Comdial	PBX	Impact Phone System	240	480
Comdial	PBX	Impression Phone System	224	192
Ericsson Enterprise Networks	PBX	Consono MD 110	26,000	26,000
Executone Information Systems	PBX	Integrated Digital System (IDS)	648	648
Fujitsu Business Communication Sys	PBX	Allegra	96	192
Fujitsu Business Communication Sys	PBX	F9600 Multimedia Platform PBX	3,000	9,600
Harris Corporation	PBX	Harris Office System	110,000	10,000
Intecom	PBX	Intecom E	16,500	Unlimited
Inter-Tel	PBX	AXXESS	512	512
Iwatsu	PBX	ADIX	200	400
KS Telecom	PBX	Atlas ISDN 500	120	460

Lucent Technologies	PBX	Merlin Legend Communications System	80	255
Lucent Technologies	PBX	Definity ProLogix Solutions	400	500
Lucent Technologies	PBX	Definity Enterprise Communications Server	4,000	25,000
Mitel Corporation	PBX	SX-200 Light	200	500
Mitel Corporation	PBX	SX-200 EL	350	500
Mitel Corporation	PBX	SX-2000 Light	500	2,112
NEC America	PBX	NEAX1000 IVS/VSP	96	256
Nitsuko America	PBX	384i	128	256
Nortel (Northern Telecom)	PBX	Meridian 1	10,000	10,000
Panasonic Telecommunication Sys	PBX	DBS 576	576	576
Picazo Communications	PBX	VS1 Business Telephone System	96	192
Redcom Laboratories	PBX	Small Business Exchange	512	512
Samsung	PBX	DCS 400si	192	336
Samsung	PBX	DCS	160	172
Siemens Business Communications Sys	PBX	Hicom 150 E Communications Server	120	250
Siemens Business Communications Sys	PBX	Hicom 300 E	22,000	20,000
Tadiran Telecommunications	PBX	Coral I, II, III	1,600	5,000
Teleco	PBX	UST DK 1000 Series	200	336
Telrad Telecommunications	PBX	Digital Key SystemBx	144	254
Teltronics	PBX	VisionLS	288	288
Toshiba	PBX	Strata DK424	200	336
TransTel Communications	PBX	Superkey SK-200	40	240
Vodavi Technology	PBX	STARPLUS Digital	96	216
Vodavi Technology	PBX	STARPLUS TRIAD	96	252
Vodavi Technology	PBX	Infinite DVX Plus	96	252
COM2001 Technologies	PCPhone	TransCom Personal Assistant	n/a	n/a

Phone System Manufacturer List

AltiGen Commuications
Fremont, CA
510-252-9712
www.altigen.com

Amtelco
McFarland, WI
608-838-4194
www.amtelco.com

Applied Voice Technology
Kirkland, WA
206-820-6000
www.appliedvoice.com

Artisoft
Cambridge, MA
617-354-0600
www.artisoft.com

AVG-Eagle
Lionville, PA
610-363-5400
www.avg.net

BBS Telecom
Austin, TX
512-328-9500
www.bbstelecom.com

Bosch Telecom, Inc.
South Plainfield, NJ
905-769-8700
bosch-telecom.de

Com2001 Technologies
Carlsbad, CA
760-431-3133
www.com2001.com

Comdial
Charlottesville, VA
804-978-2200

Computer Talk Technology
Richmond Hill, Ontario
905-882-5000
www.icescape.com

Cortelco Kellogg
Corinth, MS
601-287-5281
www.cortelcokellogg.com

Cortelco Systems, Inc.
Memphis, TN
901-365-7774
www.cortelcosystems.com

Creative Integrated Systems
Santa Ana, CA
714-513-5625
www.CISdesign.com

dbaTelecom
North Vancouver, BC
604-903-3900
www.dbatele.com

Ericsson Enterprise Networks
Research Triangle Park, NC
919-472-7000
www.enternet-us.com

ESI
Plano, TX
972-422-9700
www.esi-estech.com

Executone
Milford, CT
203-876-7600
www.executone.com

Flash Communications
Rockville, MD
301-984-8722
www.flash-comsales.com

Fujitsu Business
Communication Systems
Anaheim, CA
714-630-7721
www.fbcs.fujitsu.com

Harris Corporation,
Digital Telephone Systems
Novato, CA
415-382-5000
www.dts.harris.com

Intecom
Dallas, TX
972-855-8000
www.intecom.com

Interactive Intelligence
Indianapolis, IN
317-872-3000
www.inter-intelli.com

Inter-Tel
Chandler, AZ
602-961-9000
www.inter-tel.com

Iwatsu
Carlstadt, NJ
201-935-8580
www.iwatsu.com

KS Telecom
West Palm Beach, FL
561-840-0636
www.keysysus@aol.com

Lucent Technologies
Basking Ridge, NJ
908-582-8500
www.lucent.com

Mitel Corporation
Kanata, Ontario
613-592-2122
www.mitel.com

NBX Corporation
Andover, MA
978-749-0000
www.nbxcorp.com

NEC America
Irving, EX
972-518-5000

Newtronix
Brooklyn, NY
718-438-1888

Nexus Telecom
Kidlington, Oxford, UK
+441865847400
www.nexustelecomltd.com

Nitsuko America
Shelton, CT
203-926-5400
www.nitsuko.com

Nortel
Research Triangle Park, NC
919-992-0262
www.nortel.com

Nortel (Northern Telecom)
Nashville, TN
615-734-0000
www.nortel.com

OmniLink Communications
Lansing. MI
517-336-1800
www.omnilink-isdn.com

Panasonic
Secaucus, NJ
201-348-7000
www.panasonic.com

Panasonic Telecommunication
Systems Company
Secaucus, NJ
201-392-4220

Picazo Communications
San Jose, CA
408-383-9300
www.picazo.com

Redcom Laboratories
Victor, NY
716-924-7550
www.redcom.com

Samsung Telecommunications
Fort Lauderdale, FL
305-592-2900
www.samsungtelcom.com

Siemens Business Communication
Systems
Santa Clara, CA
408-492-2000
www.siemenscom.com

Sprint Products Group
New Century, KS
913-791-7700
www.sprintproductsgroup.com

Tadiran Telecommunications
Clearwater, FL
813-523-0000
www.tadirantele.com

Teleco
Greenville, SC
864-297-4401
www.teleco.com

Telematrix
Tamarac, FL
954-722-5905
www.telematrixusa.com

Telrad
Woodbury, NY
516-921-8300
www.telradusa.com

Teltronics
Sarasota, FL
941-753-5000
www.teltronics.com

TMC
North Brunswick, NJ
732-422-1888

Toshiba
Irvine, CA
949-583-3700
www.telecom.toshiba.com

TouchWave
Palo Alto, CA
650-843-1850
www.touchwave.com

TransTel Communications
Jupiter, FL
561-747-4466

TT Systems
Yonkers, NY
914-968-2100
www.ttsystems.com

Vodavi Technology
Scottsdale, AZ
602-443-6000
www.vodavi.com

Cabling and Wiring

Basics and Standards

Color Codes, Types and Kinds of Wire and Catagories

The best strategy for buying a communications system is to decide first on a phone system or comms server. Next, decide how and whose services you're going to use to access the outside public network(s), and then pick your equipment vendor. The last part of the buying process is to create a wiring plan in collaboration with a certified installer. If you are thinking about buying a new phone system, you should probably rewire your office.

Your existing wiring might be re-usable if it is new and/or certified. The wiriing should be clearly marked to indicate what service(s) it carries, and also should have appropriate safety labels. If the existing wiring is not marked properly you'll pay more, on average, for your phone equipment to be installed. Unmarked or illegibly marked cabling means that the equipment vendor's techs have to identify the wire by toning it out. To "tone out a cable" means that a tone generator is plugged into the jack or clipped onto a wire in the cable. The tone generator typically puts a 2 kHz audio tone on the cable under test, and the inductive amplifier of the tone testing system detects the tone and plays it through a built-in speaker. The tone generator is commonly called a "toner" and the inductive amp is called a "banana" or "wand" (because of its shape and color).

141

Copper wiring has a distance limitation. Up to the distance limitation, sometimes as short as 100 meters, station wiring should be installed "home run", which means point to point, that is, between terminal equipment (phone or desktop computer) and the switch (or server).

Residential phone installations are sometimes wired in a "daisy chain", which means wiring run from phone to phone rather than back to a central location. Daisy chain wiring can't be used for key systems or PBXs, but is used by KSUless systems.

Solid Conductor Wire

Straight conductor copper wire is solid, untwisted wire, with two wires used to make a pair. It's used for utility connections on split blocks, for cross connections in switch rooms. The plastic insulation is blue and white striped. Also called cross connect wire, this kind of utility wire is becoming rare as structured wiring systems using modular patch panels to eliminate manual connections and connection blocks.

Fiber Optics

Fiber optics are thin glass tubes that carry signals as beams of light.

In 1966, researchers developed fiber optics to carry a telephone conversation. In 1985, Bell Laboratories of AT&T sent 300,000 phone conversations over a single optical fiber.

A single strand of fiber, running free and unfettered by switching electronics, has an intrinsic bandwidth of 25 thousand gigahertz in each of the three groups of frequencies (three passbands) it can support. Just one such fiber thread can carry twice the peak hour telephone traffic in the U.S.

Fiber optic cabling does not automatically mean more bandwidth. Copper is capable of delivering high data rates (100 Mbps Ethernet), but is sensitive to the distance of the cable run. Fiber's strength is in delivering high data rates over long distances.

It is generally not cost effective to connect workstations directly to a fiber cable run. The best place to use fiber is in the system backbone, and then only if you need to be able to transmit high data rates over long distances. See the TIA/EIA-568A Standard Backbone Cabling section later in this chapter for specifics about fiber optic cable.

Coaxial Cable

Coaxial cable is so named because it includes one copper conductor that carries the signal, surrounded (after a layer of insulation) by another concentric physical channel, both running along the same axis. The outer channel serves as a ground. Many of these cables or pairs of coaxial tubes can be placed in a single outer sheathing and, with repeaters, can carry information a great distance.

A coaxial cable has great capacity and can carry large amounts of information. It is typically used to carry high-speed data or wideband analog signals (video). Coaxial cable is the kind of cable used by cable TV companies between the community antenna and end users. It is sometimes used between telephone company central offices and customer telephone poles.

Coaxial cable was invented in 1929 and first used commercially in 1941. AT&T established its first cross-continental coaxial transmission system in 1940. Coaxial cable is being used less as UTP (Unshielded Twisted Pair) signaling systems are improving and becoming more efficient.

Twisted Pair Cable

Unshielded Twisted Pair (UTP)

Unshielded Twisted Pair (UTP) cable has one or more pairs of twisted insulated copper conductors bound in a single sheath.

High-speed data create signal waves that interfere with each other. Data over 16 megabits per second should be carried on unshielded twisted wire since at such speeds, unshielded cables emit radiation which should be allowed to escape so as not to distort the data signals. The pairs of wires inside a twisted pair cable are twisted around each other to reduce crosstalk (or electromagnetic induction) between them.

Different performance levels for inside cabling are designated categories 1, 2, 3, 4, and 5.

Category 1 wire is unshielded, traditional telephone-only cable and is not used much these days.

Category 2 is unshielded, twisted pair cable acceptable for use in data applications up to 4 Mbps. It has four twisted pairs. Cat 2 is used for what cabling contractors call plain old telephone service, or POTS

Category 3, used for voice and low-speed data transmission is shielded, twisted-pair cable. Commonly used for token ring, 10 Mbps Ethernet and 10 Base-T data networks, it has four pairs with three twists per foot. Cat 3 transmission speed is specified up to 16 MHz.

Category 4 is unshielded four pair twisted-pair cable rated for up to 16 Mbps or to 20 MHz. Category 4 is no longer used because the difference in bandwidth (the space available for transmitting data) between category 3 and category 4 is not significant.

Category 5 (Cat 5) transmission characteristics are specified up to 100 MHz or 100 Mbps. It is unshielded, twisted-pair cable used in fast Ethernet and ATM applications. Twists must be maintained within 0.5 of the termination to prevent signal crossover and interference problems.

Cat 5's four pairs of wires are each twisted together at a different tightness so that no pair interferes with the another.

Cat 5 cable is downwardly compatible, which means that it can replace lower-capacity cables. Its price varies from about 6 cents to 40 cents per foot, not including installation, depending upon the brand of cable and the type of sheathing.

You do not need Category 5 cable for ordinary voice circuits but you can afford it, it wouldn't hurt, and might be a good idea to use Cat 5 UTP wiring for voice lines.

The new standard, isosynchronous Ethernet or isoE, bundles 96 voice circuits into a 10 Mbps Ethernet for simultaneous voice/LAN service on a single cable.

Shielded Twisted Pair Cable (STP)

Shielded twisted pair wire has a metal sheath around it to prevent interference. There are two basic types of shielded cable: F/UTP, or screened, (a single foil shield over all four pairs) and S/STP, in which each individual pair is shielded, and an enveloping shield covers all the pairs.

Shielded cable might be considered when you're routing datacom cable near heavy equipment, an elevator shaft, transformer or power cables, inside hospitals, airports and utilities, in a facility that's near a military base, radio station, airport or power plant. In other words, in any situation where Electromagnetic Interference (EMI) or Radio Frequency Interference (RFI) are likely to occur.

Shielded cable can keep signal out (and in) for much longer distances than UTP and accommodate higher frequencies. Shielded cables can perform better than UTP without sacrificing data integrity.

In order for a shielded system to provide the maximum benefits, it must be properly installed. This means that the shield must be properly grounded both at the wall outlet and in the telecom closet. In an improperly installed system the shield can act as an antenna attracting signals, defeating the purpose of shielding.

Color Codes and Wiring Terminations

Wiring for different uses, such as analog, digital, and LAN Ethernet, require different pair multiples that can be distinguished by their different color combinations. Each twisted pair in multiple-pair cable is uniquely color-coded.

Here's a brief introduction to the wonderful world of wire color coding. If you want to know more about the multi-colored spaghetti and the labeling and marking specs for wiring systems see the ANS/TIA/EIA-606 document.

Standard analog and digital sets require a single pair of wires (send and receive) to operate. One of the wires in the pair is referred to as "the tip" and the other is called "the ring." The simplest color scheme, used on residential station cable (called JK), uses two pairs of wire. The first pair has one green wire (tip) and one red wire (ring). The second pair has one black wire (tip) and one yellow wire (ring). One line uses the red and green pair, the second line the black and yellow pair. Some party lines use the yellow wire as ground.

Multi-pair and business wire is based on two combinations of a primary color and a secondary color. The first pair is called, for instance, white/blue, blue/white. The tip wire is the secondary color (in this case, white), with marks of the primary color (blue). The ring wire is mostly the primary color, with marks of the secondary color (blue/white). The five primary colors are blue, orange, green, brown, and slate. The five secondary colors are white, red, black, yellow, and violet.

Pairs are marked by secondary colored "binders" in groups of five. Each pair within each group uses a different primary color and each group uses a different secondary color. Each group of five pair is wrapped in binders of the secondary colors, white, red, black, yellow or violet. This allows identification of up to 25 pairs.

The next level beyond five pair is 25-pair which is called a binder group. Tip wires have colored insulation that corresponds to that of the binder group. Ring wires have colored insulation that corresponds to that of the pair.

As binder groups are assembled into larger and larger cables, the color schemes are repeated with variations that enable identification of each level of bundling. Identification of binder groups is important because services with incompatible signals should be segregated into separate binder groups.

Telecommunications cable usually come in these sizes: 1 pair, 3 conductor, 2, 4, 5, 6, 8, 12, 25, 50, 75, 100, 150, 200, 300, 400, 600, 900, 1200, 1500, 1800, 2100, 2400, 2700, 3000, 3600 and 4200 pairs.

Instal1lation
Jacks and crossconnects are designed so that the installer always punches down (attaches) the cable pairs in a standard order, from left to right: pair 1 (Blue), pair 2 (Orange), pair 3 (Green) and pair 4 (Brown). The white striped lead is usually punched down first, followed by the solid color. The jack's internal wiring connects each pair to the correct pins, according to the assignment scheme for which the jack is designed (EIA-568A, 568B, USOC, etc.).

Horizontal and riser distribution cables and patch cables are usually wired straight through end-to-end-pin 1 at one end should be connected to pin 1 at the other. (Crossover patch cables are an exception.)

> **See the wiring diagrams at the end of this chapter for a 25-pair color chart and standard modular wiring diagrams for jacks and connectors.**

TIA/EIA Wiring Standards

This section is included so that you know the basis for many of the wiring codes and standards that your installers/vendors must meet.

In 1985 the Computer Communications Industry Association (CCIA), a group of companies representing the telecommunications and computer industries, requested that the Electronic Industries Association (EIA) develop a standard for building cabling systems. The first version of the standard was published in 1991 as the EIA/TIA 568.

The subsequent standard, TIA/EIA-568A includes additional specifications for fiber optic wiring.

The TIA/EIA-568A standard as written has three basic purposes: to establish a telecommunications cabling standard that will support a multi-vendor environment, to enable the planning and installation of structured cabling system for commercial buildings and to establish minimum performance and technical criteria for various cabling systems and configurations.

The standard acknowledges that the useful life of a cabling system exceeds ten years. Having recognized the longevity, and therefore the importance of building cabling systems, the standard goes on to recommend a wiring topology, specify minimum performance requirements for cabling, and define a standard for connector and pin assignments and for cabling performance.

Topology

The TIA/EIA-568A requires that telecommunications systems be installed in a star topology. Neither bridge taps nor more than two levels of hierarchical cross-connects are allowed. (This means only one intermediate closet can be installed between the main cross-connect in the equipment room and the telecommunications closets on each floor.)

Any patch cords or jumpers at either the main cross connect field or the intermediate cross connect fields should not exceed 20 meters or 66 feet. Cabling from the communications equipment to the first termination of the voice or data ports cannot exceed 30 meters or 98 feet.

Specifications for Cabling Practices require that to avoid stretching, pulling tension should not exceed 110N (25 1bf) for 4-pair cables.

Installed bend radii in spaces with UTP terminations shall not exceed: 4 times the cable diameter for horizontal UTP cables. 10 times the cable diameter for multi-pair backbone UTP cables.

Avoid cable stress, as caused by: cable twist during pulling or installation tension caused by suspended cable runs tightly cinched cable ties

Subsystems

TIA/EIA-568A defines six cabling subsystems. The first three, building entrance facilities, equipment room design and telecommunications closet design, are described in detail in EIA/TIA-569. (This standard even gives recommended equipment room sizes and cross connect field dimensions!) The remaining three subsystems included in 568A specify backbone cabling, horizontal cabling and the work area connections.

Entrance facilities are described in 568A simply as consisting of cables, connecting hardware, protection devices and other equipment needed to connect the outside service facility to premises cabling.

Equipment rooms are defined as providing a controlled environment to house telecommunications equipment, connection hardware, splice closures, grounding and bonding facilities and protection apparatus where applicable. (Grounding and bonding requirements are described in another standard, the TIA/EIA-607.)

The primary function of a **telecommunications closet** is to contain termination and distribution of horizontal cabling. It may also house telecommunications equipment, connecting hardware and splice closures, but does not house protection or grounding facilities.

Backbone cabling provides interconnection between communications closets, equipment rooms and building entrance facilities. It includes backbone (or feeder) cables, terminations in the intermediate and main cross-connect closets, patch cords and jumpers used for backbone-to-backbone connections, terminations from the backbone cabling to the horizontal cabling subsystem and any cabling between buildings.

The standard recognizes 100 ohm UTP (Unshielded Twisted Pair) of 24 or 22 AWG for a maximum distance of 800 meters or 2625 feet for voice. Data backbone cabling is defined as 150 ohm STP (Shielded Twisted Pair) up to a distance of 90 meters or 295 feet. Multimode 52.5/125 micron fiber optic is limited to 2000 meters or 6560 feet, and single-mode 8.3/125 micron fiber is limited to 3000 meters or 9840 feet.

The TIA/EIA 568A standard recognizes that STP backbone distances are application-dependent. The 90 meter distance for STP refers to applications with a bandwidth of 20 MHz to 300 MHz.

The 90 meter distance also applies to UTP (Unshielded Twisted Pair) cabling at bandwidths of 5 to 16 MHz for Cat 3, 10 to 20 MHz for Cat 4, and 20 to 100 MHz for Cat 5.

Performance markings should be provided to show the applicable performance category. These markings do not replace safety markings.

Services with incompatible signal levels should be partitioned into separate binder groups.

Horizontal cabling extends from the work area outlet to the telecommunications closet (one per floor is recommended) and includes the horizontal cable itself, the workstation outlet, and cable terminations and cross connections in the telecommunications closets.

Four cable types are currently recognized as options for horizontal cabling, although the first, 50 ohm coaxial cable, will probably be removed from the next 568A revision. The three primary cable types are 1) solid 4-pair 100 ohm UTP cable with 24 AWG solid conductors, 2) two pair 150 ohm STP cable and 3) 2-fiber 52.5/125 micron optical fiber cable.

The distance from the information outlet to the telecommunications closet should not exceed 90 meters or 295 feet. Six meters or 20 feet is allowed for jumpers and/or cross connections in the telecommunications closet, and three meters or 10 feet is allowed in the work area between the outlet and the device.

Horizontal cables should be used with connecting hardware and patch cords (or jumpers) of the same performance category or higher. Installed UTP cabling shall be classified by the least performing component in the link.

It is acceptable to specify category 3 wiring for the telephones and category 5 for the data. Category 3 is rated for speeds up to 10 megabits per second, and category 5 up to 100-155 megabits.

Wiring can be brought to the jack in many ways: either by in-wall wiring, or by floor wiring concealed by plastic wiring covers, conduit under floors

The **work area** should have a minimum of two outlets, one for voice and one for data. The voice outlet should have one 100 ohm UTP four pair cable for voice, wired with either T568A or T568B.

The data outlet can be wired with 100 ohm UTP four pair, 150 ohm STP two pair, or 62.5/125 micron fiber.

These specs on unshielded twisted-pair cabling supersede TSB-36 and TSB40-A.

Horizontal UTP Cable in the work space is specified as solid 4-pair 0.5 mm (24 AWG). Solid 0.63 mm (22 AWG) is also allowed. An overall shield is optional.

Components and installation practices are subject to all applicable building and safety codes that may be in effect.

Performance marking should be provided to show the applicable performance category. These markings do not replace safety markings.

UTP Connecting Hardware

Most outlets, connectors and jacks used in cabling systems are specified by TIA/EIA.

Transmission requirements are much more severe than for cable of a corresponding category.

Performance markings should be provided to show the applicable transmission category and should be visible during installation (for example CAT 5) in addition to safety markings.

Connecting hardware should be installed to ensure a well-organized installation with cable management, and in accordance with manufacturer guidelines.

Outlets shall be securely mounted. Outlet boxes with unterminated cables must be covered and marked.

Installed connectors shall be protected from physical damage and moisture, and be exposed to a temperature range no greater than -10∞C (14∞F) to 60∞C (140∞F).

Connector Terminations

Installers should strip back only as much jacket as is required to terminate individual pairs. Pair twists shall be maintained as close as possible to the point of termination, untwisting shall not exceed 25 mm (1.0 in.) for category 4 connections and 13 mm (0.5 in.) for category 5 connections.

UTP Patch Cords and Cross-connect Jumpers

Patch cords must use stranded cable for adequate flex-life.

Must meet minimum performance requirements for horizontal cable except that 20 percent more attenuation is allowed for stranded cables.

Insulated O.D. of stranded wires should be 0.8 mm (0.032 in.) to 1 mm (0.039 in.) to fit into a modular plug.

Performance markings should be provided to show the applicable transmission category in addition to safety markings.

Performance specifications for plug cord assemblies are under study.

More Specifics, and Some Advice

Jacks

Six-wire RJ14 jacks are used for telephone service, RJ45 for 10BaseT and 100BaseT service. Leave a spare UTP (unshielded twisted pair) cable hanging loose behind the modular jack panel for future use.

Single line phones, accessories, answering machines, and modems use the RJ11C or RJ11W jack. Two line phones, accessories and answering machines use the RJ14C or RJ14W jack. Three line phones and accessories use the RJ25C jack. Four line phones and accessories use the RJ61X jack. Burglar and fire alarms circuits use the RJ31X or RJ38X jack. Single line fixed loss loop data installations use the RJ41S or RJ45S jack. Four wire data circuits use the RJ48C/RJ48X or RJ48S jack. One to 25 single or multiple line circuits bridged to the network or customer equipment use the RJ21X. RJ22 is the designation for the smaller jack that is used on handsets.

A common configuration is a quad (four outlet) faceplate with one jack for the telephone, one spare for fax or modem, one jack for data and a fourth location with a spare data jack. All of this initial capacity, in addition to accommodating future technological growth, will allow you to squeeze desks for holiday help in with your regular employees without adding more cable.

Flush mounted jacks typically have single and double (two connectors or four) backboxes. Backboxes are the five-sided empty boxes that go into the wall. They're generally at the end of a metal conduit that runs inside the wall, and act as a switching housing for the wires and connectors.

Faceplates and More

Single- or double- gang faceplates go over the open side of the gang box flush on the wall and provide service outlets to individual desktops and workstations. Look for modular, snap-in connectors (the new rage) and sockets (RJ-11, coax, RJ-45, etc.) you can mix and match to your heart's content. Also look for a reliable termination system (good insulation-displacement connectors [ICD], and the like) and good-quality exterior connectors. Angled connectors (modular plugs that aren't perpendicular to the plate) require less desktop clearance. Look for ease of installation-you'll have to install a lot of these components. Easy installation will give clear benefits in terms of time and materials, and the fact that the goodies are installed will increase overall system reliability.

Cable Pairs

Generally speaking, you want to provide as many pairs of cable to each desk or workspace as possible. A realistic cabling installation uses two-four pair category 5 cables to each location.

Twisted-pair network connections usually require two pair. Telephones need from one to four pair, depending upon the technology. Each modem or fax connection requires a pair of wires.

Keep an Eye On

Patch panels and rack systems are used for hooking up the near ends of cable in the switch room. (the far ends go to your desktops, etc.) Look for good labeling schemes, some built-in wire-management routing and some indication on front-panels that connections are live. On rack-mountable components look for front-mounted connectors. On all equipment demand good-quality, corrosion-resistant modular connectors.

Connecting blocks/split blocks are used mostly for attaching telephone trunks to CPE (Customer Premises Equipment), or hooking up voice cable home runs on the house side of a switch. Look for a tough wall-mounting system, integral covers and label slots, good-quality metal contacts (these tend to corrode over the years, and get a lot of wear from repeated punchdowns), well-designed Insulation-Displacement Connectors (IDCs: these are designed to cut through insulation on single wires, making a firm contact), built-in wire-bundling, integration of blocks with standard connectors (e.g., a split block, prewired to a 25-pair AMP connector can make life easier in certain situations).

Raceways, ducting, grommets, etc. conceal and protect wires as they travel through and around walls, baseboards, and through holes in desktops. The best systems allow convenient installation and provide removable covers for quick repairs after installation.

Labeling systems Some are generic, some purpose-matched with lines of equipment. Both help you mark, and hence keep track of, wires and connections. The most advanced systems integrate a physical label-management system (a way of sticking labels on stuff and making sure they're protected), and a logical management system-some way of facilitating record-keeping, complying with labeling standards, and establishing a feedback loop into cable-management software. Top of the line systems may even incorporate portable label printers that accept downloads from cable-management software and generate labels in the field.

Installation tools and test equipment A very mixed bag. Remember that a tool that really works is cheap at any price, so aim for quality and pick the right tool for the job.

Structured wiring management software lets you build a graphic map of the facility so that you can locate and maintain an inventory of the cable in your walls and the equipment that connects to it.

Demarcation Point
Also called the demarc. The demarcation point is the area between the wiring that comes in form your local telephone company and

the wiring you install to hook up your own telephone system-your CPE (Customer Provided or Premises Equipment) wiring. A demarc might be a as simple as an RJ-11C jack (one trunk) or an RJ-14C (two trunks) or an RJ-21X (up to 25 trunks). If you are ordering your own network services, ask your equipment vendor how the telco service should be installed. Some systems terminate with modular plugs, some are direct-wired to a telco mirror block.

Short note about wiring and the demarc: It is a good idea to have your lines installed on an RJ21X punch-down block rather than on jacks with modular connections. It's a good idea to wire a phone system into an intermediate "mirror block," rather than directly onto the telephone company's RJ21X or lightning protection block. The mirror block is a duplicate of the RJ21X, but is owned by you, as customer premises equipment. If there are any problems with your trunks, they can be safely tested at the intermediate point. Any pulling or poking at wiring into the RJ21X could cause wires to touch together and short out, or dislodge wires from their connection. It's best to leave the telephone company's connections alone.

A note about lightning protection: The phone company is required to install lightning protection between the incoming lines and your equipment, but often doesn't. Be sure to ask specifically for lightning protection to be installed. The demarcation point is located on the subscriber's (customer's) side of the telephone company's protection equipment.

Troubleshooting

Wiremap Tests

A wiremap test checks all of the wires in a cable for these transmission errors:

An **open** is a lack of continuity between pins at either end of a cable.

A **short** means that two or more lines are short-circuited together.

A **crossed pair** is connected to different pins at each end; for example pair 1 is connected to pins four and five at one end, and pins one and two at the other.

A **reversed pair**'s two wires are connected to opposite pins at each end of the cable, for example the wire on pin 1 is connected to pin 2 at the other end and the wire on pin 2 is connected to line. A reversed pair is also called a polarity reversal or tip-and-ring reversal.

A **split pair** has one wire from each of two pairs connected as if it were a pair, for example: the Blue/White and White/Orange lines connected to pins four and five, White/Blue and Orange/White to pins three and six.

NEXT (Near End Crosstalk)
The result of a split pair is excessive Near End Crosstalk (NEXT). NEXT wastes 10Base-T bandwidth and usually prevents a 16 Mbps token-ring from working at all.

A cable with inadequate immunity to NEXT can couple so much of the signal being transmitted back onto the receive pair (or pairs) that incoming signals are unintelligible.

Cable and connecting hardware installed using poor practices can have their NEXT performance reduced by as much as a whole cable category.

Attenuation
A signal traveling on a cable becomes weaker the further it travels. Each interconnection also reduces its strength. At some point the signal becomes too weak for the network hardware to interpret reliably. Particularly at higher frequencies (10 MHz and up) UTP cable attenuates signals much sooner than does co-axial or shield-

ed twisted pair cable. Knowing the attenuation (and NEXT) of a link allows you to determine whether it will function for a particular access method, and how much margin is available to accommodate increased losses due to temperature changes, aging, etc.

Other Cable Tests

Cable length is usually checked using a time domain reflectometer (TDR), which transmits a pulse down the cable, and measures the elapsed time until it receives a reflection from the far end of the cable. Each type of cable transmits signals at something less than the speed of light. This factor is called the nominal velocity of propagation (NVP), expressed as a decimal fraction of the speed of light. (UTP has an NVP of approximately 0.59-0.65). From the elapsed time and the NVP, the TDR calculates the cable's length. A TDR may be a special-purpose unit such as the Tektronix 1503, or may be built into a handheld cable tester.

Testing for Impulse Noise The 10Base-T standard defines limits for the voltage and number of occurrences/minute of impulse noise occurring in several frequency ranges. Many handheld cable testers can test for this.

Cabling and Wiring Manufacturers, Distributors and Publications

AMP
Harrisburg, PA 17105-3608
1-800-722-1111
1-800-245-4356 (Faxback service, USA)
(905) 470-4425 Canada
(617) 270-3774 (Faxback service, Canada)

Anixter (international cable products distributor)
see Anixter 199x Cabling Systems Catalog
Anixter, Inc
4711 Golf Road
Skokie, IL 60076
(708) 677-2600
1-800-323-8167 USA
1-800-361-0250 Canada

32-3-457-3570 Europe
44-81-561-8118 UK
65-756-7011 Singapore

AT&T Canada
Network Cables Div
1255 route Transcanadienne
Dorval, QC H3P 2V4, Canada
(514) 421-8213
Fax: (514) 421-8224

AT&T documents
AT&T Customer Information Center
Order Entry
2855 N. Franklin Road
Indianapolis, IN 46219 USA
(800) 432-6600 (USA)
(800) 255-1242 (Canada)
(317) 352-8557 (International)
Fax: (317) 352-8484

Belden Wire & Cable
POB 1980
Richmond, IN 47375
(317) 983-5200

Bell Communications Research
(Bellcore)
60 New England Ave
Piscataway, NJ 08854
(800) 521-2673
Fax: (908) 336-2559

Berk-Tek
(copper and fiber optic cable)
312 White Oak Rd
New Holland, PA 17557
(717) 354-6200
1-800-BERK-TEK
Fax: (717) 354-7944

Black Box Inc.
(connectors, switch boxes and outlets)
P.O. Box 12800
Pittsburgh, PA 15241
1-800-552-6816 (USA)
(412) 746-5500
Tech Support USA
(416) 736-8013

Tech Support Canada
info@blackbox.com

CableTalk
(racks & physical cable management)
18 Chelsea Lane
Brampton, ON L6T 3Y4 Canada
(800) 267-7282
(905) 791-9123
Fax: (905) 791-9126

Cabling Business Magazine
12035 Shiloh Road, Ste 350
Dallas, TX 75228
(214) 328-1717
Fax: (214) 319-6077

Cabling Installation
& Maintenance Magazine
One Technology Park Drive
POB 992
Westford, MA 01886
(508) 692-0700
Fax: (918) 832-9295
Subscriptions: (918) 832-9349

Comm/Scope Inc.
POB 1729
Hickory, NC 28603
(800) 982-1708 (USA)
(704) 324-2200
Fax: (704) 328-3400

Corning Optical Fiber Information Center
1-800-525-2524
Guidelines is their publication/
newsletter on fiber technology.
fiber@corning.com

Graybar
(international cable products distributor)
1-800-825-5517
(519) 576-4050 in Ontario, Canada
Fax: (519) 576-2402

Hubbell Premise Wiring Inc. (cable
and structured wiring manufacturer)
14 Lords Hill Rd
Stonington, CT 06378
(203) 535-8326
Fax: (203) 535-8328

The Siemon Co
(cabling system supplier)
76 Westbury Park Rd
Watertown, CT 06795
(203) 274-2523
Fax: (203) 945-4225

MOD-TAP
(cable and equipment supplier)
285 Ayer Rd
P.O. Box 706
Harvard, MA 01451
(508) 772-5630
Fax: (508) 772-2011

Northern Telecom
(cable and physical network products)
Business Networks Division
105 Boulevard Laurentien
St. Laurent, QC H4N 2M3 Canada
(514) 744-8693, 1-800-262-9334
Fax: (514) 744-8644

Ortronics (structured wiring systems
and wiring products)
595 Greenhaven Rd
Pawcatuck, CT 06379
(203) 599-1760
Fax: (203) 599-1774

Saunders Telecom
(racks, tray and accessories)
8520 Wellsford Place
Santa Fe Springs, CA 90670-2226
(800) 927-3595
Fax: (310) 698-6510

Siecor
(structured wiring manufacturer)
489 Siecor Park
P.O. Box 489
Hickory, NC 28603-0489
(704) 327-5000
Fax: (704) 327-5973

Structured Wiring Systems

A structured wiring system is an integrated set of connectors, jacks, patch panels (constructed, installed, and labeled the same way), often packaged with software-running-on-a-computer, that organizes, labels and tracks your wiring and connections.

Structured wiring systems are standards-compliant. A system built with purpose-matched components can be trusted to maintain transmission, radiation, building code compliance, and other standards, when hooked together with appropriate cabling or fiber.

Equipment product cycles are getting shorter; and new standards, better equipment and faster processors make replacing your technology every one to three years economically desirable. Wiring is considered part of a building's physical plant, and its useful life is typically ten years. Your cabling will have to work through several equipment-upgrade and network topology cycles.

A well planned and scrupulously maintained structured wiring system helps you adapt to inevitable changes in technology. Knowing what you have in your walls and ceilings lets you make new connections and perform adds moves and changes faster and more reliably. Knowledge of your wiring system facilitates long-term planning. You can make better decisions about the equipment to buy based on what you know about your network capacity, strengths and weaknesses.

Without a structured, modular system for planning and delivering multiple voice and data services, you're setting yourself up for higher installation and maintenance costs, end-user frustration, and less-than-efficient service.

A structured wiring system should let you understand your in-house network so that you can use this knowledge for better planning and maintenance. If all your wiring information is in one place, it is easier to process and monitor your installations and adds, moves, and changes.

Structured wiring standards make it unnecessary merely to predict how your system will perform. If you can maintain and keep the standards you've set, your network will be running at its best. You'll have fewer problems due to bad connections or faulty wire, and if you do have a wiring failure it will be simple to locate and troubleshoot.

Installation is simplified. Your installers need learn only one basic system for installing panels, bundling wires, establishing clearances, testing, labeling, etc. Most systems have specialized tools that simplify the installer's job. A specific tool can save days of labor over the course of a large installation.

Structured wiring system designers have encountered and designed around most of the typical wiring problems of an installation. A good system will offer solutions for tight quarters, sharp corner bends, odd clearance, interference conditions, multiple conductor types, shallow walls, etc.

Structured wiring systems often include management software. You keep track of cables, wires, jacks and extension assignments using a PC-based cabling management package.

See the *Telemanagement and Cable Management Systems* chapter for more about cabling management systems.

Cable management software is often included with call accounting software. See the *Call Accounting* chapter in this book for more.

Telemanagement and Cable Management Systems

PC-based software for organizing and tracking your phone system and structured wiring system

Good telemanagement means staying on top of your telecom equipment and costs so that you can collect and disburse relevant information in a timely way and in a truly useful format.

Telemanagement packages are software-based, and include integrated applications you can add depending on your needs. Here are the modules typically found in a telemanagement software bundle:

For more information about call accounting systems, read that chapter in this book.

Call Accounting
Call accounting lets you track and bill your telecommunications costs, and identify and prevent fraud.

Integrated Directory
The directory lists who is assigned to which extension, and also keeps information on equipment assignments, email addresses,

and which employee is in what department. Attendants and receptionists can look up employees to transfer calls or answer questions about how to reach them.

Inventory Management

Stay on top of your equipment inventory. Proper inventory management allows you to avoid duplicate purchases and keep often-used equipment in stock.

Work Orders

What equipment is out for repairs? When's it due back? Is this a good repair company or are they expensive and slow? Sometimes this software handles moves, adds and changes (MACS) within the company, too.

Network Analysis

Checks the cost-effectiveness of your call routing across the networks. This often extends to Internet use, and some software even tracks where employees go on the Internet.

Cable Management

Keep track of your wiring. Which desks are wired for data as well as for phones? Which desks have second lines for faxes and modems?

A cable management system (CMS) is a structured relational database in which each desktop outlet is clearly tied to a physical location, an employee, a wiring conduit, and a patch panel or termination-also (ideally) a vendor, a cost, an installer, and a maintenance history.

A CMS keeps track of cables, their types, specs and connections. It also keeps track of cable runs, racks, cross-connects, and patch panels, outlets and other parts of the active physical cabling plant. This information is supplemented with a library of cable, connector, and other hardware specifications used in maintenance and also the costing of new construction.

The CMS correlates all of the above with departments, functions, and users, and with the physical layout of your space. It frequently employs built-in graphics and/or provides schematic output to architectural drafting software.

Finally, a good CMS will document procedures for handling repairs, moves, adds, changes, testing, and will act as a dispatch system for work orders and trouble tickets.

While all these functions are crucial, procedures, and the way they are handled, are the most important aspect of a CMS. The most important thing you can do with a CMS is to keep it updated. Cable plants have a way of migrating out of true with record-keeping systems, eventually reaching a point where records no longer reflect reality. To avoid this, a CMS must enforce updating of its databases before any change to actual wiring can be considered complete.

Parts definitions may be loose, based on abstract cable and equipment types and overall characteristics, or they may be tight, based on actual vendor part numbers and specs. There are benefits to the latter approach, but in general, it requires more interdepartmental cooperation and effort to maintain a system that tracks numbers and specs.

Once the CMS is fully loaded with your company's information it should become your dispatch center for all work done on the cabling system. Some specialized CMSs work with custom patch panels and special cables to create a partially automated system. The specialized systems report cable status through the data network and they manage and document moves and changes with minimal direct human intervention.

Whichever modules you decide you need, a good telemanagement package must have an underlying relational database so that the data you enter can be accessible to all the software applications in the package. You don't want to re-enter data separately for the directory, call reports, inventory, etc.

It's also very helpful if the database is ODBC-compliant. ODBC (Open Database Connectivity) is a standard used by many software programs including Excel and Microsoft Word. ODBC lets you import/export data in/out of your favorite formats so you can view the information the way you like it, without having to learn new software from the ground up.

Once your telemanagement software is in place, you can cut overall communications costs. Automate some functions (for instance, shutting down access to your PBX at night to fight toll fraud or scheduling call accounting reports to run automatically). Reduce redundant staff (or reorganize more efficiently), and track equipment so it doesn't "walk off."

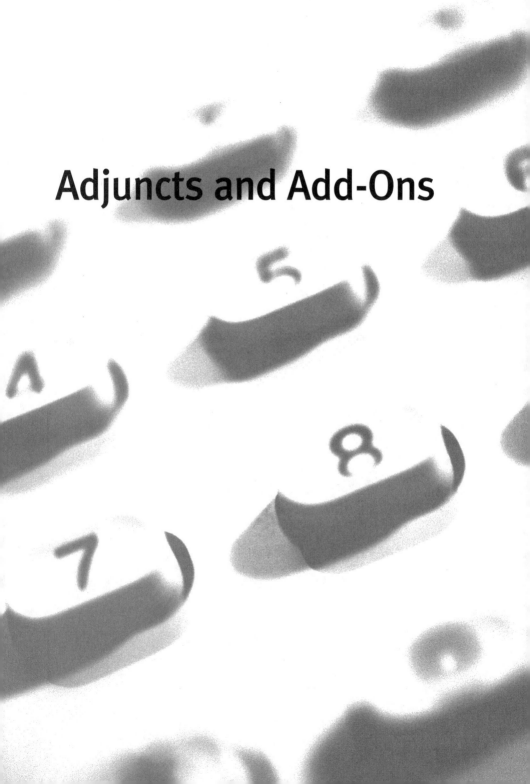

Adjuncts and Add-Ons

Fax

Fax, infrequently called telecopying, is the transmission of scanned-in or computer-based text and images to a telephone number associated with an output device. The information is transmitted digitally as electrical signals through the telephone network. A receiving fax device (machine, server, PC card) reconverts the coded image and stores and/or prints a paper copy of the document.

Fax on demand and fax broadcasting are powerful tools for boosting sales and are relatively inexpensive. Fax on demand gives your customers and prospects the ability to call your voicemail system anytime and request fax documents. It's an inexpensive, labor-saving way to disseminate product brochures and technical support notes and FOD can capture data, turning callers into prospects. Fax broadcasting is essentially telemarketing via fax-you're sending one fax document to a large group of possible customers.

Faxes can be sent to and from network-based fax servers, voicemail systems, PC-based fax modems and fax service-bureaus. LCR (Least Costing Routing) schemes over intranets and the Internet and DID inbound fax routing are the latest technologies to look at. Fax identification and routing is today's biggest technological condrum.

Technical Basics

The standards promulgated by the ITU (International Telecommunications Union) are called Recommendations and the recommendations for fax are the T series which govern the fax protocols and the V series which govern modem operation.

The most widely known industry standards are CAS (Communicating Applications Standard) invented by Intel and others and tied closely to the Intel architecture, and FaxBios (developed by an industry consortium) which is less machine-dependent.

Fax machine standards describe how the machines work. These standards are divided into groups. The T series ITU recommendations apply to Group II and III fax standards. Group I faxing predates the Group II (T.3) and Group III (T.30) recommendation.

Groups I and II
Groups I and II fax are obsolete standards. Group I fax machines take about six minutes to send a page and have a resolution of 98 scan lines/inch. Group I devices work by attaching the document to be transmitted to a drum rotating at 180 rpm along which a photocell moves. The paper on the drum makes loud flapping noises if it is not attached securely and transmission takes nearly forever, which, in the early days, made faxing very expensive. Group II fax (the ITU's T.3 standard) transmits a page in about three minutes at a resolution of 100 scan lines/inch.

Group III
Group III fax machines are the most-installed and purchased machines. They have a less-than-one-minute per page transmission speed (top speed is 9600 bps) and the standard resolution is 203 by 98 pixels (dots) per inch (dpi).

Many Group III (which is the ITU's T.30 standard) fax machines use non-standard ways to get higher resolutions (300 x 300 dpi and 400 x 400 dpi).machines. Extensions to the Group III standard to support these higher resolutions have been proposed, but to get these resolutions you probably should have the same type of machine at both ends of the transmission.

Two fax machines must negotiate a common resolution, page width, and page length before sending each page. All Group III fax machines have to support the standard resolution and A4 paper size so that common ground can always be found.

The basic coding scheme for Group III fax is called one-dimensional coding (also known as MH or Modified Huffman), in which each scan line of pixels is compressed. Group III also supports two-dimensional (MR or Modified READ) compression, but this method of compression is computationally intensive (so it takes a long time or a fast processor) and most (inexpensive) fax machines do not support it.

Group IV

Group IV is the standard for digital fax transmission over ISDN at a speed of 64 Kbps. (This is one B channel or half the capacity available on a BRI [Basic Rate Interface] connection). There's lots more information about ISDN in the Provisioning section of this book.

Group IV faxes transmit one page in around three seconds. Standard resolution is 400 x 400 pixels per inch and requires an improved compression scheme.

T.6 is the ITU recommendation that covers the image compression algorithm used for Group IV fax machines.

OCR and ICR

Documents that require modification should be sent via email and not faxed unless absolutely necessary. Email files are sent in ASCII text or HTML format so they are not images and can be edited without image-to-text conversion.

The contents of a fax (text or images) are treated as a single graphic image and converted into a bitmap in the process of faxing. This means that letters and characters don't mean anything to a word processing program so you can't change a faxed document that you receive on your computer without converting it with OCR (Optical Character Recognition) or ICR (Intelligent Character Recognition). Fax as a graphic image also means that fax servers (more about routing for fax servers below) can't understand where the faxes it receives are delivered, because the cover sheet looks like one big picture and can't be read.

A OCR or ICR software package converts images into text by scanning the image and matching letter shapes to something it knows. A fax server can scan the cover sheet text for the name of the fax recipient and route the fax using OCR and ICR.

OCR is generally not very reliable, though, especially if the faxes are not sent with fine resolution, and the faxes and cover sheets can't be handwritten because the OCR isn't smart enough to read it. Handwritten cover pages will wind up being manually routed. Some fancy fonts frequently elude OCR applications, too.

ICR (Intelligent Character Recognition) is a better, but more expensive, method to convert fax images to text so that recepients can be identified.

The difference between ICR and OCR is that ICR uses more powerful (but much slower) algorithms so it can read standard resolution faxes. ICR is still not entirely reliable, but it is an improvement over OCR.

Other Fax Server Routing Technologies
OCR and ICR are better than nothing, but not as good as DID routing (more about DID later).

T.30 sub-addressing is another way to route faxes. It works as long as the sender's hardware can specify the sub-address when the fax is sent. Sub-addressing is a new feature added to the T.30 protocol that permits the sending fax machine to specify where the fax is to be delivered. The user enters the fax routing information after entering the fax phone number and a # (pound sign).

Most standard fax machines don't support sub-addressing so it is not a very useful method of identification.

Bar-Code Routing

The sender of the fax uses a bar code on the cover sheet to represent the desired recipient of the fax.

The sender has to be able to generate bar codes and be willing to place the stickers on the faxes or be willing to use cover sheets provided by recipient. This routing method is covered by a patent and not generally available without special licensing.

Transmitting (Calling) Station Identification Routing (CSID)

CSID is the originating fax name/number, usually printed across the top of the fax. CSID routing is used when all faxes from one source go to a single destination. Fax servers relate the CSID text with a destination fax mailbox box or machine.

Faxes from the machine using CSID are not permitted to go to any other destination than the one specified, so this routing method works if the two devices talk only to each other.

Received Fax Line Routing

The fax server has several trunks attached, one for each fax line and each fax modem. All the faxes received on a designated fax line are routed to a particular output device. A separate phone line and fax modem is required for each user who wishes to receive faxes, so this routing method tends to be expensive and can't be cost justified. If you're going to pay for a separate line for every device, you might as well not have a fax server at all.

DID Routing

The DID (direct inward dial, provided by local telco) method is the best and most reliable routing method for inbound faxes. You can use DID service with various fax recipient devices, but it is something in particular to look for in a fax server. It takes some special hardware as well as one or more dedicated inbound DID fax lines.

The phone company rents DID phone numbers to you in blocks of 20 or 25 (usually) and also charges you (slightly more than standard business line rates) monthly for the trunks. There are some setup and installation costs.

Each person gets an individual extension, but the calls are routed by the telephone company over one or just a few trunks. How many DID trunks you'll need for fax routing depends on the traffic. If people are receiving only short faxes sporadically, a couple of DID trunks can easily handle 100 users.

If a fax is sent to any of the numbers in the DID block you've rented, the CO (Central Office) sends down an assigned trunk. The CO sends the particular number that was dialed across the DID trunk using DTMF signals before it starts the call that tells the device on the other end which of the DID extensions was dialed.

The downside to DID trunks is that they are inbound-only and can't be used for outbound calls. You must use standard analog phone lines or analog ports off of your phone system for outbound fax calls. The upside is the reliable and secure automation of fax routing. You can assign each of your users a private DID fax number. When faxes come in, they're routed down to that specific workstation. It's nice and neat.

Not all fax hardware boards support DID routing so make sure that your fax server can handle DID if you want to use it.

Fax Applications

Fax on Demand

Fax on Demand (FOD) is an often-used Computer Telephony (CT) application. It combines PC-based fax components and CT processing software; a FOD or faxback system involves both voice and fax processing. Callers ring into the system. The computer answers and asks "What documents would you like? Here's a menu." Selections are made through touchtone responses and the FOD computer sends the faxes selected. FOD usually works with voicemail systems.

There are two types of FOD systems. The first is a one-call machine. The caller rings in from his own fax machine. When he's chosen his faxes and ready to receive, he simply hits the fax's "Start" button and gets the information. Here, the caller pays entirely for the phone call needed for the FOD transaction, unless it's provided on a toll-free number.

The second type of FOD system is called two-call. The caller phones in and, besides telling the system what documents to fax, he also inputs a fax number where he wants the documents sent. The supplier of FOD must pick up the toll charges on the second leg of the transaction because the system makes a second, outbound call to send the faxes.

You could also break FOD down even further by distinguishing between faxtext FOD and Interactive Fax Response.

Faxtext FOD sends each caller identical documents (for example movie schedules, spec sheets, product brochures). Interactive Fax Response is similar to Interactive Voice Response (IVR) because individual callers get customized information based on who they are and what, using a touchtone pad, they tell the machine to do.

FOD always offers two prime benefits to the user of the technology: employees don't have to waste time faxing documents over and over again (it's automated); and customers learn to love it because it's an easy way for them to get information.

FOD Features to Look For

One- and two-call operation A few years ago, it was hard to find software that handled both. Today, you can and should be able do both. One-call operation saves the sender (you) money and it gets the information to the caller faster. Two-call FOD lets the machine send outbound when the fax line or lines are free (and cheaper). You can also receive more requests for information per line because the trunks are not tied up sending information during busy inbound traffic periods.

International call blocking The system should allow you to block (or allow) expensive international calls.

Blocked number list The system should support a list of numbers where faxes can't be sent in two-call operation. This can save money by reducing crank calls.

Live operator transfer During specified business hours, a caller should be able to break out of the FOD system and get to a human. This should be a supervised transfer so that the caller doesn't get RNA (Ring No Answer) and busy signals. The transfer codes should be configurable for different PBXs.

Menu tree structure Better systems support a menu tree where applications can be secluded from each other. For example, two separate FOD information lines could be supported from the same system, with document numbers private to each line. Menu trees are also good if two different companies use the same FOD system.

A related feature is DID support for inbound calling that allows the FOD system to handle multiple FOD applications. DID makes a lot of sense for FOD service bureaus or companies with different divisions using the same system. (See DID Routing in this chapter.)

Index page generation The better FOD systems support automatic index document generation. Unless you want to upload a custom one, the system can build a standard catalog page from document numbers and descriptions fed into the system when each individual document is uploaded.

Binary File Transfer (BFT) New FOD systems should include the option of Binary File Transfer for delivering files to users with fax modems. BFT adds another level of functionality to FOD by letting you send spreadsheets, software updates and other binary files to callers. You can send software updates and instructions on how to install it from the same system.

Local and long distance detection As area codes are split, telecommunications systems need to be able to handle more difficult dialing plans. The FOD should be able to tell which a local call and which isn't.

Flexible fax loading There are several ways to get the documents you're sending into the FOD system. This is an important feature because it's one that the actual user will have to handle. The simplest method is by just using the fax machine as it stands. The quality of fax documents that have been scanned in is usually poor. A better way is to give PC users a special Windows print driver on their machines so that they can directly print to a graphics format file from any Windows program. Then the fax is ready to send over a network to the FOD system directly from the desktop.

Online updating This is nice when you have to change a document or some other part of the FOD application and don't want to take the FOD application offline. Some of the better applications can remain online during document updates. Others require you to wait until all callers have hung up before changing anything. Waiting can be a real time-waster when the FOD system is busy.

Other Useful Features Fax support on voice mail systems for inbound calls means that fax messages can be delivered to voice-mail boxes when a message arrives or during specific hours determined by the owner. If the faxes need to be changed at the desktop, Optical Character Recognition (OCR) or some other recognition technology (See the Fax Servers section in this chapter for other recognition methods) is used to translate the faxes to text.

The text can be run through a text-to-speech (TTS) program, which creates an audio message for the recepient. Faxes can also be sent to the owner's mailbox and stored there and re-faxed directly to the owner's computer or fax machine at a later time.

Outbound fax over voicemail works with FOD applications. Callers into the FOD system can request that faxes be sent to them.

Fax Broadcasting

A fax broadcasting machine, many times folded within the same PC that handles FOD, automates the task of bulk faxing. You tell it what needs to be sent, give it a list of fax numbers and it fires everything off, usually during off-peak hours to cut down on toll charges.

Fax Broadcasting Features to Look For
Simple loading Make sure it's easy for people to load documents into the application.

Smart re-transmission When faxes are interrupted, it's good to have a system that will re-establish transmission and fax only the remaining pages of the interrupted fax, rather than sending the whole document again. This is a particularly useful feature if you send large documents.

Public and private lists The fax broadcasting application should support easily modifiable and uploadable lists for system-wide and private use. Some systems let you maintain a text file, then upload it at the time of the broadcast.

Be wary of fax databases Fax numbers get changed much faster than phone numbers. Lists of fax numbers always seem out of date. Spend time updating your lists so they remain current.

Least Cost Routing (LCR) Better systems offer LCR. You use either specific lines for calls to particular areas or dial prefix codes to select a long distance carrier based on the time of day and the destination phone number.

Cover Page Customization The fax broadcasting application should let you build a cover page with fields where text can be imported at the time of delivery. This could be a name, routing instructions and possibly a personalized message.

Even better is a system that lets you customize each of your bulk faxes. This is a great way to avoid being labeled a "junk" faxer.

Multi-document Broadcasts You should be able to send multiple documents with the same call, since a longer call is cheaper than several smaller ones. Many broadcast systems won't let you send more than one document at a time.

Broadcast Cancellation You should be able to cancel a broadcast at any point and create a new list of those people who didn't get the original fax.

Delayed Broadcasts The system should let you schedule your broadcast overnight for cheaper long distance rates. If the transmission takes more than one night to complete, it should suspend the broadcast until the next evening.

Priorities You should be able to set a higher or lower priority on each broadcast. A lower priority will make sure the normal functionality of the system is not affected by your broadcast. A broadcast with a higher priority can take over the system until it is complete.

Status Logging You need to know how far down a broadcast list the broadcast has gone.

Voice-annotation of Fax Messages You can add a spoken comment to faxes that you send or forward. The fax-broadcast (or fax-on-demand) application can convert WAV files recorded on a PC to voice prompts.

Fax Servers

A fax server is a computer that sits on a LAN and has one or more PC fax boards in its expansion slots. The fax server receives incoming faxes over phone lines, stores them on its hard disk and, if it knows who the fax is for, alerts that person over the LAN. If it can't identify the fax, it can print it or store it in a central depository that individuals (or a supervisor) can check manually.

On the other end of the client/server connection, a fax server accepts faxes from workstations on the LAN, stores them and gets them ready for sending. It can send the faxes immediately or wait until later, when phone calls are cheaper. Fax servers generally handle bulk faxing tasks, meaning your fax server can be a fax broadcasting engine too.

Fax servers make faxing a lot easier and more cost effective than having a bank of standalone fax machines that people have to trundle over to when they want to send a fax. Fax servers are environmentally sound, too, since the need for paper is reduced. Most fax servers log transmissions and produce reports, making fax costs easier to track. And fax servers are certainly a cheaper solution than giving every workstation a modem and an analog phone line.

LAN fax servers have a sticky technological drawback, however. How can they recognize the recipient of an incoming fax in order to route the fax appropriately?

One of the problems with computer-based facsimile that was never anticipated by the developers of the original fax standards, is the need to be able to route incoming faxes to the appropriate user on a network. Usually faxes arrive at a fax machine with no electronic addressing information and only a typed or handwritten name on the cover sheet.

If you don't solve the routing problem, your faxes to a fax server will probably have to be printed and manually routed. This means a human looks at cover pages, figures out who gets what and then puts the faxes in people's boxes-a process that is not secure, reliable or labor-saving.

Fax Server Features to Look For

Email Integration This is becoming a necessity. Make sure the fax server software interfaces properly with email platforms so that faxes can be translated into email messages for off-site retrieval without a fax machine.

Ease of Use End users can cost your compay more money in setup and training time than the system actually costs. Make sure sending and receiving are simple to do. Some companies offer downloadable demos on their websites so you can try before you buy.

Image Resolution Make sure your server creates the fax images using the same resolution that it uses to transmit the fax. Some fax server products scan images at 300 dpi (dots per inch) and then scale them to 200 dpi during transmission. This can cause a noticeable loss of clarity.

Instant group support Better solutions let you select any number of users from a user database and add them to an "instant group" as the fax destination. Otherwise you have to build a list from scratch and enter each one individually.

Workstation Digitization This feature takes the digitization workload from the server and puts it on the desktop. The PC converts the file that is to be sent into a fax image file that is then sent to the fax server.

The workstation must have enough diskspace or access to network disk space to create a fax image file. Workstation digitization also causes more data to be sent through the network so dialup users will see very slow performance.

Server Digitization This means that the server has access to many different applications and can take a file, open the application that it was created in, and create a fax image. This lowers the file size being sent from the user over the network, since most files are smaller before they are digitized into fax images. The downside is that the fax server must know how to translate each type of file that it will be asked to send.

Remote administration Get as much capability for network and remote administration as possible so that your vendor and/or inhouse support people can get to and solve problems wherever they are.

Phone number normalization The system should be able to determine if the fax number specified is a local, toll, long distance or international number. If the number is not prefixed with a 1 and needs it, the system should add the prefix automatically.

Real-time status monitoring (versus the traditional store-and-forward method) lets you send and monitor the progress of faxes immediately.

Caller ID Systems that support caller ID let you track ads for responses. Summary statistics based on caller ID let you know who's buying or receiving your documents, which are the most popular, and what's working.

Fax over the Internet

Several Internet-based services accept email messages and fax them to the specified phone number. Some are free and most require that you have email or Web access. Some require that you install special software on your computer.

Other commercial services include bulk faxing, and web page form input to fax.

Services that allow faxing from the desktop let you to send a fax using software on your Internet-connected computer. Some services simply send your fax from one central Internet server, others use the Internet to partially route the fax to save long distance charges. It works something like this: faxes are transmitted from the sender's desktop fax software to a special server. The fax server routes faxes via an Internet connection to a remote server located in geographical proximity to the destination fax machine. From this remote server, the fax is sent to its target address using the local telephone system.

Fax Modems

Most modems are capable of sending and receiving fax data. Fax/modem software generates fax signals directly from disk files or the screen.

Fax modems can also transmit data. A standard proposed by the EIA/TIA/ANSI, called Binary File Transfer (BFT), extends the fax Group III modulation and protocols for bulk data transfer. The problem is that few fax modems or software packages provide support for this mechanism.

Fax Switches

A number of devices on the market try to distinguish between an incoming voice, fax, or data call and route the call appropriately.

Fax switches attach to the phone line on one side and on the other side to the other devices (your phone/answering machine, fax machine or modem).

All devices work on one of two general principles, either listening for fax tones or voice, or listening for distinctive ring patterns (cadences).

In the first case the device will answer the phone and try to guess what it should do based on what it hears. Some machines play back a sound of a phone ringing so that humans dialling in think the phone is still ringing when in fact the fax switch is listening to see if the call is from a fax machine or a human. If the tone from the calling fax machine is heard, then the switch connects the call to the fax machine, otherwise the call is deemed to be a voice call and is connected to your phone/answering machine.

A slightly more sophisticated approach is for the fax switch to answer the phone and play a short recorded announcement. If, during the announcement the fax tone is heard, then the call is switched to the fax machine. If no tone is heard but sound is heard after the announcement, then the call is assumed to be voice and switched appropriately. If nothing is heard then the switch either considers the call a data call and switches it to a modem or considers it a fax call from a machine that does not generate the right tone and switches it to the fax machine.

The other approach relies upon an optional service available from some telcos called SmartRing, Distinctive Ring, RingMaster or Ident-a-Ring. This feature allows you to have more than one phone number associated with the same phone line. Incoming calls using the different phone numbers can be differentiated by the different ringing patterns The fax switch distributes the call based on the ring cadence it detects.

The advantage of the first (switch) approach is that you don't have to pay the phone company or depend upon the feature being available. The disadvantage is that it is not always reliable, especially when fax machines that do not generate the required special tones are sending the incoming message.

The advantage of the second (telco) approach is that it is very very reliable. It requires the availability of the SmartRing feature from the telephone company and involves an additional monthly charge.

Fax Modem Standards

At the hardware level, the two standards that govern the exchange of commands between a host computer and a fax modem are EIA-578 (Class 1) and EIA-592 (Class 2).

At the software level there is one official standard, the ITU-T T.611, plus a number of industry standards.

Class 1

The Class 1 fax modem standard describes an extension to the "Hayes Modem Command Set" that permits computers to send and receive faxes using fax modems. The Class 1 standard is a low-level specification in which most of the protocol work (T.30) and image generation (rasterising [see glossary] and T.4 compression) is done by the computer (in software) while the modem handles only the basic modulation and the conversion of the asynchronous data from the computer into the synchronous packets used in fax communications.

The primary advantage of Class 1 modems is that fax protocol is implemented in software which means that additions and changes to the standard can be implemented without requiring a hardware modification.

The primary disadvantages are that the software vendor has to handle the complexity of the T.30 protocol and that Class 1 is very sensitive to timing and some operating systems, and has great difficulty meeting timing constraints and maintaining the fax connection. Removal of the timing limitation is the primary reason for the proposed Class 4 standard.

Class 2

The Class 2 fax modem standard describes another extension to the Hayes Modem Command Set to permit computers to send and receive faxes using fax modems. The Class 2 standard is a higher-level specification; most of the protocol work (defined by ITU's T.30 recommendation) is done by the modem while the computer is responsible for managing the session and providing the image data in the appropriate format (This format is covered in the T.4 recommendation).

The primary advantage of Class 2 is that the low-level detail work is handled by the modem. Not only does this mean that software developers do not have to support the T.30 protocol, it also relieves the host computer of all of the time-critical aspects of fax communications.

The Class 2 standard took a long time to be approved (more for political than technical reasons) and many companies did not wait for the final version to be approved before shipping modems. As a result most Class 2 modems adhere to the first draft and not to the standard as it was approved. To compensate for this, the approved Class 2 is referred to as Class 2.0 and the draft version as plain Class 2. Be careful when you buy.

Class 3

Class 3 is reserved for a future project to define a standard for fax modems that would, in addition to handling the T.30 protocol

(Class 2), also handle the conversion of ASCII data streams into images (T.4). Although there are a couple of fax modems that handle the ASCII-to-fax conversion, no draft document has been circulated, and the future of this project is in doubt.

Class 4

Class 4 is Class 1 with intelligent buffering to reduce the need for the host computer to respond instantly to the fax modem. There is a draft of this standard but no official recommendation at this writing.

Fax Service Bureaus

Fax services are available for a low monthly fee plus per-page charges or at per-minute rates. Fax service providers will do your fax broadcasting, can handle your fax overflow, and provide fax-on-demand and fax mail services. High volume faxing can be customized so that every recipient receives a personalized fax. Call the phone company for phone lines, and service bureaus for faxes, if you don't want to do it yourself.

Faxing From Phone Systems

If you want to add fax messaging, fax mail, fax broadcast, or fax-on-demand services to your phone system, use fax add-ons that your phone system manufacturer may offer. System phone users can retrieve faxes by using deskset LCDs and softkeys. Fax answering by the phone system's auto attendant and fax overflow (storing faxes when fax lines emit busies) are other nice features that Nortel, for one, offers as an option when you buy the Norstar system.

Fax Legaleze

(This information is for reference only.)
The Telephone Consumer Protection Act (TCPA) is a federal statute enacted on December 20, 1991. The TCPA restricts the use of fax machines to send unsolicited advertisements. The

TCPA also directs the FCC to adopt regulations to protect residential users' right to refuse to receive telephone solicitations to which they object.

An "unsolicited advertisement" is defined as a transmission advertising the commercial availability or quality of property, goods or services without the prior express invitation or permission of the person or entity receiving the transmission.

FCC rules require that each transmission to a fax machine contain, in a margin at the top or bottom of each transmitted page or on the first page of the transmission, (1) the date and time the transmission is sent (2) the identity of the sender and (3) the telephone number of the sender or of the sending machine.

(It is important to note that the requirement to mark faxes with this identifying information applied to fax machines and not to fax cards used in computers. This loophole is shut, pending reconsideration proceedings.)

The person on whose behalf a facsimile transmission is sent will ultimately be held liable for violations of the TCPA or FCC rules.

The TCPA and the FCC rules do not preempt a state law which imposes more restrictive requirements or regulations. You should contact the state public utilities commission in the state you're working in to find out what laws apply to your business.

Voicemail

Voicemail systems record, store and let you retrieve voice messages. The technology has been around long enough that most manufacturers feature similar offerings-you can pick up messages, forward messages to other mailboxes (with comments), leave messages, edit messages and deliver messages to a mailbox at a prearranged time. Messages can be tagged "urgent" or "non-urgent" and stored for future listening. Most voicemail systems also provide auto attendant functions: they'll pick up incoming calls, play a series of choices and greetings, and then transfer the caller to the appropriate extension or mailbox.

> **Auto attendants are covered in detail, with IVR (Interactive Voice Response), in the next chapter.**

A basic voicemail box works like an answering machine. It records messages when you are unavailable. It also lets you send messages from your voicemail box to other voicemail users, assign your own personal passwords, confirm message receipt, send messages for future delivery, and maintain day/time information on messages. Basic boxes should store a flexible number of administrator-defined messages.

An enhanced voicemail box allows subscribers to receive faxes in their boxes, maintain group distribution lists, and establish guest mailboxes that non-subscribers can access.

An announce-only or listen-only mailbox communicates a message without allowing callers to leave one of their own. These boxes are used to disseminate repetitive information to callers.

ECP, or Enhanced Call Processing, mailboxes are most often placed at the beginning of an IVR application. Callers are allowed to access menu items through their telephone keypad.

Transcriber mailboxes and Forms mailboxes work in tandem. The Forms box is programmed to ask the caller a series of questions, allowing time for responses. The Transcriber Mailbox records each of the responses. These mailboxes allow people to make a phone call instead of hand-writing a form and returning it by mail.

A fax station mailbox is assigned to an extension associated with a fax machine. When an incoming fax encounters a busy or unanswered machine, the fax station box will receive and store the fax, then attempt to deliver it to the fax-printing destination.

Fax Broadcast mailboxes are also associated with a particular fax telephone line. Fax broadcast mailboxes have group distribution lists linking other fax machines and fax-ready mailboxes. They are used to send a single fax to many destination numbers.

A voicemail system communicates with your telephone system by sending analog signals such as flash hooks and touch tone (DTMF) sounds. Digital telephone systems use proprietary digital signaling between the extensions. To connect an analog device, like voicemail, to a digital telephone system requires an analog station card inside the system or a special digital link between the voicemail system and your phone system. Digital links work better than analog connections. Digital links between phone systems and voicemail systems are usually available only with a large PBX and a voicemail system made by the same manufacturer, or as part and parcel of a computer-based phone system or communications server.

How to Choose a Voicemail System

A voicemail system can be one of the best tools a small business or home office can have. It can improve the service you give your customers, get sales information to prospects more quickly, give a more professional image to your company, and enable you and your busy staff to work more efficiently.

The wrong system can be a nightmare to you and to your customers. Keep your voicemail system customer-friendly. Before you buy a system consider the impact on your customers. Will they be better served? Will you maintain friendly, human service in spite of the voicemail system?

Find out if your existing or proposed phone system will support voicemail or an automated attendant. Voicemail/auto attendant applications require that your telephone system must be able to transfer calls from one extension to another.

True call transfer capability means that you can move a call from one telephone extension to another. If you have to put the call on hold, buzz another extension, and tell them to pick up the held line then you do not have true call transfer capability.

A voicemail system must perform "call progress detection" for reliable transfers from the phone system to the voicemail system. The voicemail system listens for rings, busy signals, speech and drops in line current during a call transfer. Systems without call progress detection won't integrate reliably with your telephone system and call transfers might not always happen the way they should.

When selecting a voicemail system, consider how many simultaneous phone calls need to be handled by the voicemail. To handle more than one phone call at a time (either coming into the system or between inside extensions), you need a multi-line (or port) voicemail system. If you're using auto attendant you'll need more ports, because there will be several calls in the system at one time.

When selecting a voicemail system, decide how much labor will be required by the vendor to install the system. Labor is expensive. Think about the responsibility your company will assume. Do you need employee training or will someone in-house learn how to use it and train everybody else? Is there someone who will provide the vendor with a list of employees, extensions and features or will you pay to have them get the information? Will someone in-house make ongoing changes?

Don't buy a standalone system unless you have to. They don't integrate with your phone system, so you won't get an auto attendant or message notification. Your phone won't be able to transfer calls from your extension automatically to your voicemail box, either. Standalone systems are connected directly to your telephone trunks, and are really useful only as an answering machine substitute.

How To Size A Voicemail System

Voicemail systems are sold by the port, starting at two ports and expanding to 24 or more. The average telephone system with 100 people usually calls for an 8 to 12 port system. This means that 8 to 12 people, either coming into the voicemail system from outside or accessing the system from inside, can use the system at the same time. The number of ports a voicemail system offers (and can expand to) indicates its size, and loosely, its price.

Some voicemail systems have "dynamic" ports, which can be used either for automated attendant or for voicemail. If your system has dynamic ports, consider how many additional calls may need to be answered simultaneously by the automated attendant. Each of these calls is using a port while the call is being switched to an extension or department through the auto attendant. Once the caller reaches the extension, the port is freed up.

Another thing to consider when sizing a voicemail system is the number of hours of storage. Voicemail systems use software, processing power and system memory in different ways, so it is important to understand your particular system.

Here's what to tell the vendor so that you get the voicemail system that is the best size for your company:

The number of users you have, including your remote users. The length of the messages that the users will leave. (Should it be unlimited or do you want to set a finite time limit for the callers?) Also tell the vendor how long the messages will be stored, and the growth in the number of users you anticipate over the next three to five years.

Voicemail Alternatives

Telco Voicemail

Another kind of voicemail is the service purchased from thelocal telephone company or cellular provider. It costs from $5 to $15 per month.

Most digital cellular service providers provide visual message-waiting notification. When you have voicemail messages waiting in your mailbox, a voicemail symbol appears on your telephone display. No need to call in constantly to check for messages. If the icon appears, you know you have new messages.

Most phone companies provide a voltage pulse (to flash an LED) and/or a stutter dialtone to notify you when you have messages.

Single-line SOHO offices and heavy users of cellular telephones can benefit from voicemail boxes provided by the telco. You won't miss important calls. Don't, however, use a telco-provided voice-mail behind a PBX or key system; you will be much better served by a dedicated, integrated voicemail system.

Voice/Data/Fax Modems

These are low-cost modems with voicemail capability that are installed in desktop PCs. They're the least useful type of voice-mail service. Before you buy a voice/data/fax modem, consider these points:

Voice/data/fax boards are modems with a voice chip, not full-blown voicemail systems. They have limitations. They are not designed to give high quality sound or to conserve disk space; true voicemail systems use cards designed specifically for voice-mail. You'll get better sound quality with dedicated voicemail solutions, and true voicemail systems use compression algorithms to minimize the disk space required for messages.

The software included with a voice/data/fax modem is limited. You give a caller a specific menu that transfers the caller to a mail-box to leave a message. No other options are available. With a true voicemail system, you can set up as many menus as you like, nest menus within menus, and give the caller options to move around the system. Voice/data/fax modems do not perform call progress detection and can't be used to transfer calls reliably.

Voice/data/fax modems can work well as an after hours answering machine for a single incoming line. If you'd like more flexibility, consider a dedicated voicemail system.

Faxing From A Voicemail System

Fax capability integrated into voicemail lets you offer Fax on

Demand (FOD) to callers. FOD is a great way to disseminate printed material automatically.

Check the Fax chapter in this book for more about FOD.

Most voicemail cards—especially multiline cards—do not have fax functions built in. You will need to add both a fax board and additional software to your voicemail system to add fax capability.

Installing and Using A Voicemail System

Voicemail can help your business give better service to customers and potential customers or, used improperly, it can drive them away. Voicemail is a tool. It can free you from answering repetitive questions and let you better manage your time, but it has to be used responsibly and responsively.

Voicemail box holders can improve the way the system works for your company. Boxes should be checked frequently, at least twice a day.

Boxholders should delete old messages when they are no longer useful. This will free up voicemail system storage space.

Tell your boxholders to record a greeting using their own voices. Using default system greetings gives the impression that you don't take the time to check or return messages.

Users should update their greetings for vacation days and other absences. Some voicemail systems have an alternate greeting feature so that you don't have to change your regular greeting every time you leave and return to the office.

Boxholders should be careful when recording greetings and always listen to the recording after it's made. It is helpful to include the key that callers can press to skip the greeting next time they call and get your box. Most systems use either # or 1 for that purpose.

Ask for the following features when you request voicemail proposals from vendors.

Voicemail Box Features

Simplified mailbox activation. First time users should be able to setup a mailbox quickly and easily by recording their name, message and entering a security code.

Message retrieve, save, copy and delete. Users should be able to retrieve messages in their mailbox internally or off site, save a message for future reference, copy a message to other mailboxes or delete messages that are no longer needed.

Certified, priority and private messages. Certified messages give the sender notification that the message has been received and listened to by the intended receiver. A priority label should be available for messages so that the message can be heard first. The same system assures that private messages can't be sent elsewhere.

Distribution lists. This feature allows a list of mailboxes to be created and assigned a distribution list number. One message can be sent to the entire group. Each mailbox should have the capacity for several lists and an overlapping of mailboxes in broadcast lists should be allowed

Fast-forward/rewind, message skip. Users can fast-forward or rewind messages for a preprogrammed, flexible length of time. New and saved messages can be skipped. It should be possible to skip new messages and listen to saved messages.

Date and time stamping. Messages include a system announcement indicating the date and time the message was sent. This can be programmed to be heard by the user at all times or can be programmed only to play if the user requests the information during playback of the message.

On/off site programming and recording. Mailbox users can change mailbox features and settings using a touch tone phone either onsite or offsite.

Password protection. Mailboxes should have an optional security code so that messages cannot be heard unless the code is entered.

Offsite notification/Enhanced notification. The system should be able to alert users to mailbox messages by dialing out to their pagers, cellular or home phones. Enhanced notification allows several telephone numbers to be dialed, with a flexible number of attempts and at user-selectable intervals.

Messages left by Name. Callers should be able to leave messages for mailbox holders by spelling part or all of their last name.

Immediate Reply. Your voicemail system should allow users to record replies to internal messages immediately, without entering an extension or mailbox number.

Context-sensitive help. As they are using the system, mailbox users and outside callers should be able to get feature explanations that are specific to the task they are trying to accomplish.

Multiple greetings. Users should be able to prerecord up to five mailbox greetings.

Call Processing Features

Caller holding/queuing. When callers are transferred to a box-holder extension, and if that extension is busy, the caller can be put on hold and reminder messages played to inform the caller of their status in the queue. The timing parameters between reminder messages should be programmable.

Call screening. Boxholders should be able to tell who is calling. The voicemail system prompts callers to record their names and then plays the name to the boxholder. The called party can decide whether to accept the call or have it sent to their voicemail box.

Definable primary/secondary operator. Most systems can assign a specific extension that the incoming calls will forward to if the caller presses 0. Some systems allow designation of more than one extension in case the first is unavailable.

Dial-through during greeting. Callers should be able to dial another extension while the mailbox greeting is being played.

Fax tone detection. Your voicemail system should be able to detect incoming faxes and send them to an extension with a fax machine attached. If the extension is busy, the voicemail system should be able to store the fax until a machine is available.

Administrative Features

Full maintenance and reporting. The administrator should have access to complete system maintenance and programming, and should be able to view or print reports.

Extension/mailbox creation and editing. The system administrator should be able to add or edit extensions and mailboxes at any time.

Remote maintenance and modem. Complete system programming, diagnosis and maintenance should be accessible remotely through modem and system software.

Automatic disk optimization. During designated system clean-up time, the hard drive should be defragmented and compressed. Deleted and old information should be removed during a programmable time period.

System back-up/restore. A system programming backup should be recorded in case it needs to be used to restore the voicemail system. System backup should be automatic at a preset time.

System password protection. Access to system maintenance and programming should require security code.

Summary

If possible, you should purchase a voicemail system at the same time you buy a new telephone system. The price will be lower because the technicians are already installing the phone system and the incremental time to install voicemail is negligible. Buying your voicemail and phone system from the same vendor may decrease your options, but it will simplify the purchase and installation. You'll be able to negotiate discounts and maintenance contracts on both systems at one time, and more importantly, you can design the way the system handles calls from end to end.

Interactive Voice Response (IVR) and Automated Attendants

Auto attendant and IVR systems enable callers to execute certain transactions online without the intervention of customer service people. Typically, 30 to 60% of inbound calls are repetitive, so IVR saves money on labor costs because it gives information to callers automatically. IVR also serves your customers better, because information is accessible 24 hours a day, plus, if you get information from callers your data will be better; IVR interfaces reduce input errors.

Auto Attendants

Automated attendants are generally packaged with voicemail systems. Auto attendants answer your incoming calls directly and present the caller with choices that can be accessed using touch-tone signaling.

Auto attendants replace a live person answering the phone. You free an employee from answering all of the incoming calls, and your callers can help themselves by transferring to mailboxes that play recorded announcements or by connecting directly to an extension.

The way an auto attendant is designed determines whether or not the system will work for you. It is crucial that design can make or break your auto attendant implementation. Your callers need to be presented with a clear, easy to understand menu of choices, and a way to "opt out" to a human, if they must.

Look for these features when buying an auto attendant:

Transaction processing
Callers should have 1 digit access to departments or to specific announce-only mailboxes.

Flexible channel assignment and time settings
It should be possible to record separate greetings for day, night, weekends and holidays, and the system should automatically switch to the appropriate greeting.

Voicemail/extension directories
The system should allow a caller to dial by name and then be transferred to the appropriate extension or directly to a specific mailbox.

Interactive Voice Response

Interactive Voice Response "gives data a voice". By understanding the touch tone pad on the telephone or the spoken voice (when IVR is combined with speech recognition), IVR applications can understand commands from callers and lets them request, manipulate, and in some cases, modify data that resides in a database on another computer.

IVR usually works in front of an ACD, so that calls can be transferred from the IVR system to customer service if the caller so chooses.

IVR systems perform functions including receiving calls, playing a menu of choices, recognizing DTMF tones or a spoken command, playing the requested information (that's either an audio file or data in ASCII format), verifying a caller's identity, retrieving/updating information (stored on the host PC or the company's LAN), sending a fax, and ending the calls.

IVR applications can be combined with text-to-speech technology, which reads ASCII-stored information out loud to callers.

The technology is "Interactive" because the user is prompted for information by the IVR system and the IVR system returns specific information or performs certain tasks based on the input from the user.

IVR is a computer telephony application; it's usually PC-based, and interfaces to outside voice circuits or to a PBX or key system. The actual database from which the information is retrieved is a separate product and sold by a different vendor.

Accessing a database is the key difference between auto attendants and IVR systems.

Synthesized voice, also called Text-To-Speech (TTS), is used for reading ASCII information, usually information like numbers and dates. These systems interpret printed information, translate it to speech and read the information to callers.

IVR can be integrated with fax technology, and then the IVR system can include faxed confirmations of the transactions and retrieval of pre-stored documents (called Fax on Demand).

Calling number ID recognition by an IVR system saves the caller the step of identifying him/herself via a touch tone pad and allows associated database records to be retrieved even before the voice greeting is played. Caller ID also allows calls to be intelligently routed to departments, messages, or individuals.

IVR systems are generally used in larger operations, where customers helping themselves make a big difference to the bottom line. Most small to medium-sized businesses can get the same functionality from an auto attendant/voicemail system.

How to Design an Auto Attendant or IVR Script

Most of your callers will not remember more than three or four options in a list of choices. They will certainly be bored and frustrated by a series of menus that keep them from obtaining the information that they are looking for. If your callers are inbound on your toll-free number, every costly, billable word is tying up your inbound facilities, system hardware and storage unnecessarily.

Get to know your audience. If they will be using the application often, you might to include a "power user" option at the greeting level. Gear your presentation to the type of person that you expect to use your application the most, but don't forget the other callers.

Really understand your application. What must callers know or do at a particular point in time? What are they looking for? Try to be really specific here, it's very easy to try to tell them everything, or to assume that everyone knows something and leave out a key detail or two.

Any time and money that is spent in pre-production and planning will be returned in the quality and consistency of the finished production, and in your profit margin. It's not ready until it's tested.

Scripts should be read to others and the listeners should be asked to react. The best test is to create the application and test the script by getting people from outside your industry to test it in action. It is all too common to read a complex script silently only to find that it does not hold up in the final recording or that something was forgotten that has to be recorded later, at additional expense.

It's what the caller hears that counts. Be consistent. Always use the same style of presentation in your instructions. Be frugal with words. Always present the option first and then the associated touch-tone number to press.

Keep the menus short. Use "focus groups" for complex scripts. Focus groups are people who try your application and then are interviewed to find the good points and the flaws before the application goes online.

Look at your script, and your audience very carefully. Professional talent that will do the best job for you probably isn't as expensive as you think. In New York, for example, the talent for a 100 prompt script with approximately 15 minutes of material costs, on average, between $350 and $600.

Never assume. Describe, to all involved parties, in detail, just what your application is, what you need to accomplish and what is expected from each of them. Since IVR and auto attendant prompt recording is a relatively new field, never assume that the talent or the studio has ever recorded prompts for an IVR system before. Even if they have, don't assume that they understand your particular application and its particular audience and requirements.

Call Accounting

Call accounting systems give a nice, neat package of information that helps you analyze employee performance, guard against toll fraud and best configure your telephone system. A handful of employees can make thousands of dollars of improper calls a month, and extra trunks that aren't being used cost money, too. On the other hand, too few phone lines incoming phone lines can cost you customers and business.

Call accounting usually is an in-house function, a software package running on a dedicated PC that connects to the phone system. Accounting services are also available from an outside company (called a Service Bureau) that charges a fee per call ($0.01-$0.05 cents per call depending upon call volume).

Call accounting systems are easy to set up, easy to use, and well worth the investment. You can buy basic call accounting software from lots of vendors, and install it on a PC you buy separately. Your telephone system vendor probably offers pre-packaged systems that they will hook up for you.

Call accounting systems use SMDR (Station Message Detail Recording) information from your phone system and manipulate the data so that it means something to you.

Your phone system spits a continuous stream of SMDR data as calls are made, and in some cases, received. The SMDR port, usually on the backplane of the system's KSU, feeds these calls, on-the-fly, either directly into your onsite PC-based system or into a memory buffer/polling device where they are stored until they're collected from off-site.

Each record in the data stream represents one phone call. The data record is a series of numbers and letters that give information about that particular call, including the start and stop times of the call, which extension originated the call, what number was dialed and which trunk the call used to exit your phone system to the PSTN (Public Switched Telephone Network).

Each manufacturer's key systems and PBXs format this information differently, so it's important that your call accounting system can read your phone system's particular SMDR record coding scheme.

Incoming call data may also be available, typically for calls over T-1 circuits and into toll-free numbers. This information lets you see which trunk was used, where the call came from, which extension took the call, if it was transferred, and if so, to where and how long the call lasted.

Call accounting software organizes call information for you and prepares reports, matching the call detail with data that you've provided about your employees and phone system.

Understand that call accounting systems (whether they're on- or off-site) don't come ready to go. You have to provide information about your company so that the call data makes sense to you. And the information has to be kept current so that as employee extensions, your phone system trunk configurations and your local and long distance service pricing change, the call accounting system can get the correct call records from the phone system and can see which extension placed a call, can match the name, department, division and cost center to the extension number, and tell you who is using that extension. And so that when the phone system lists the trunk number that was used to send the call out of your system, the call accounting system matches the trunk number to the service that's being used.

The duration of the call, when referenced against your rate tables, is used to tell you how much each call costs your business.

The best-known and most-used reason for installing a call accounting system is to prevent fraud. Extension usage reports tell you who is calling where and for how long. It is possible to set thresholds in the call accounting software so that when a certain call length or call cost is exceeded, the system will alert you, either by flagging the call or extension in a report, or, depending upon the severity of the excess, will give an on-screen warning, an email or will beep you. You can get reports organized by enterprise, location, division, department, and extension, and disseminate the reports to the appropriate section manager.

You can generate reports on demand or schedule them by day, week or month. Usually the systems are set up so that the reports are generated and printed (or sent to you via email) at an off-peak time.

Other Things a Good Call Accounting System Can Do

Monitor Trunks

You can find out how and when calls are going over each of your phone lines. If all of the trunks are busy a lot of the time, you probably need more. Every time all of your lines are busy you are missing calls. Conversely, if you have trunks that never get calls, they're either not necessary or not working.

Allocate Phone Costs Among Sites or Departments

Costs can be separated according to your company's products or divisions, so you can easily see where the money goes. The costs and usage of hardware and facilities are easy to divide and assign back to departments.

Allow Sharing and Resale of Long Distance and Local Phone Calls

In businesses like hotels/motels or hospitals, separate bills for each room must be generated. With a call accounting system, you can be your own phone company.

Evaluate And Motivate Personnel

Check your data to see how well your sales or customer service reps are doing. Find out how long it takes them to wind up a call, how many calls, on average, it takes for them to make a sale, and where customers are coming from. Use these results to improve your telemarketing plan.

Network Optimization

Figure out the best combination of carriers and rates/services, and which is the best combination of all the various services each offers.

Verify Long Distance Bills

The bill from your carrier isn't always accurate. Call accounting systems double-check your long distance, itemizing each call so you can tell if and where you were mischarged.

Trace Calls

Sophisticated systems let you see what customers are calling in on what lines. This way you can check the effectiveness of telemarketing campaigns.

Import Non-PBX Call Data

Calling card and cellular phones use can be imported and merged into your phone system's call data so you can report all calls made by an employee or business unit, not just the inside calls.

Rate Tables and Call Costing

Theoretically, a call accounting system, when loaded with information about what you pay for phone service, can tell you how much each call costs in real time. But it is not that easy. The biggest problem is getting accurate information, and the second is that the call accounting system might not be able to use the information you get.

You'll probably have to input your local calling costs manually. It's a lot of work. Long distance companies might give you a copy of your rate table on disk, but it's not very likely. Vendors of call accounting systems can provide basic rate tables for the biggest carriers, but they probably aren't the same plans or prices that you've negotiated, assuming that the vendor-provided rates are even current.

Many local and long distance calling plans are based on volume. Call accounting systems usually aren't sophisticated enough to keep cumulative minutes-used totals, and, when the next price break occurs, they certainly can't go back and re-cost a call that has already been made and reported.

Real-time call costing numbers should be used as a short-term reference, not as a basis for chargeback. If you assign telecom costs to departments or individuals, best wait for the bill. It is possible to get your bill from the phone company electronically; on disk, CD, or via a special communications link.

Service Bureaus

Service bureaus are a nice alternative to owning your own system because you don't have to buy, install, maintain and update your own system. You get nice, neat, reports and departmental invoices delivered the way you want them, and can assume that changes you ask to be made, are made.

Today, service bureau call accounting is hot, especially for larger businesses. The Internet has revolutionized the way a call accounting service bureau can deliver your information.

Service bureaus can deliver your reports and invoices via the Web, with myriad access alternatives. Each person in your organization can get an individual, customized report delivered to his/her desktop.

Call accounting service bureaus collect SMDR info from a buffer box installed next to your switch. Buffer boxes are cheap devices filled with short-term memory (the more memory they have, the more expensive they are) that connect to your system's SMDR port and to an outside telephone line. The service bureau dials into the box over the phone line and downloads the data. Bigger companies often use the same kind of buffer box and polling setup for collecting data from remote offices to process call data centrally.

Summary

Unless your business is so small that you can track phone usage by eyeballing your bills, you should buy a call accounting system. The software (and PC) are inexpensive enough that they'll pay for themselves almost immediately.

Employees, even if you don't hold them accountable for personal calls, tend to stop making inappropriate calls once they know their calls are being reported. Most companies see a 10% to 30% reduction in personal calls just by installing a system.

Power Protection

Irregularities in electrical power, if strong enough, can damage telephone and computer equipment. Sometimes the damage is not immediately apparent. Smaller inconsistencies can, over time, cause electronic component damage, data corruption, signaling and processing errors.

You need power protection: surge protectors to guard against spikes, power conditioners to keep the flow of electricity to your telephone system and computers at a constant level and wave-form. You should have an uninterruptible power supply (UPS), so that when your power goes out, you have time to shut down your equipment, save files if you need to, and wind things down so that they can be restarted and be expected to work.

A power surge is an over-voltage, or increase in voltage lasting at least one cycle or 1/60 of a second. Surges can occur when equipment that is drawing large amounts of power is removed from the circuit. Surges can also be caused by utility company load management or by discharges of static electricity.

Spikes are short bursts of high voltage, typically from 200 to 6000 volts and are commonly caused by lightning strikes. Also called impulses, spikes can be caused when utility power comes back on line after having been lost.

Noise is evidence of electromagnetic interference (EMI) or radio frequency interference (RFI). Electrical noise may be intermittent or chronic, and can be caused by many factors, including lightning, load switching, and radio transmitters.

Brownouts are long-term, multi-cycle under-voltages lasting minutes or even hours. Brownouts are often deliberately introduced by utility companies when peak demand exceeds generating capacity and can also be caused by ground faults and sudden start-up of large electric loads.

Sags are temporary, short-term drops in power supply. Sags can cause equipment malfunctions and hardware damage.

Surge Suppressors

Surge suppressors are devices that plug in between your equipment and the commercial AC power outlet, and clamp or reroute excess current. When a surge occurs on the power lines, the surge suppressor sends the overload to ground.

Important features to look for in surge suppressors are:

Clamping level
The level of increased voltage at which the surge suppressor kicks in; the lower the clamping level, the better the protection.

Let-through voltage
The amount of voltage that gets through to your equipment at a given incoming voltage level. The lowest let-through rating UL tests for is 330. The standard UL 1449 rating records voltage when a 6,000-volt transient spike hits.

Response time
The length of time from the onset of a surge to when the protection equipment "clamps."

Power Conditioning

Power conditioning makes sure that sags (temporary drops), spikes, (brief, sudden surges), or brownouts (lengthy reduced current) don't reach your machines, and also smooth out AC waveform into the consistent shape that microprocessors prefer.

A power conditioner combines a voltage regulating transformer and an isolation transformer. It provides smooth, regulated, noise-free AC voltage. Know the range of output voltage when shopping. Some products are designed to feed conditioned power at a constant 120 VAC (Volts Alternating Current), others let output voltage vary.

Uninterruptable Power Supplies

Uninterruptable Power Supplies (or UPS systems) are used to provide a redundant (backup) power source for critical data and voice equipment. When UPS systems are combined with power protection technology, the quality of the AC power supplied to the equipment is improved as well.

Redundant power means that the equipment to be protected is provided with two power sources, a primary source and a backup source in case the primary source fails.

Power defects that can be improved by a UPS include surges, spikes, noise and sags.

There are two standard UPS technologies, standby or on-line. Standby systems use the existing AC power as the primary power source to the equipment. The batteries and DC-to-AC inverter together act as the backup power source. When power from the utility company fails, a transfer mechanism switches the load to the batteries and inverter. When the batteries/inverter is put into use, the inverter acts as a filter for the backup power.

An on-line system uses the battery/inverter as the primary power source and the existing input AC power as the backup source.

The difference in power topologies is most obvious when there is an AC input power failure. In a standby power system when the primary AC power provided by the electric company fails, the load transfer is made from the input AC to the battery/inverter source. On-line power systems do not have to transfer the electrical load when there is an AC failure because the input AC is not the primary source, the batteries and inverter are.

On-line power backup systems eliminate the possibility that data and communications connections will be disrupted in the case of an AC power failure, but do not eliminate load transfers from primary (battery) to secondary (input AC) power. Primary power failures in on-line systems occur if the batteries or inverter fail or if the inverter's power load changes suddenly.

Standby and on-line power backup systems require different size battery chargers. The battery charger in an on-line system must be capable of handling the full electrical load required by the equipment. A standby charger can be much smaller because it needs only to keep the backup batteries charged while they are in standby.

A hybrid topology called line-interactive power backup utilizes a reversible inverter that is always connected to the output of the UPS. When the AC input power fails, the power flow is from the battery to the UPS output. When the input AC power is normal, the inverter reverses and charges the batteries. The inverter also improves the reliability of the raw input AC by filtering and regulating the power stream. Standby power systems use input AC power as the primary power source without an inverter, therefore the primary power to the equipment is not conditioned.

How to Size an Uninterruptible Power System (UPS)

Here's a step-by-step calculation that will help you find the UPS that's best for you.

1. List all equipment that will be powered by the UPS. Include all monitors, terminals, external hard drives, hubs, modems, routers and any other critical equipment.

2. List the amps and volts. These figures are listed on a plate on the back of the equipment.

3. Multiply amps by volts to determine VoltAmps (VA). Some equipment may have power requirements listed in watts (W). To convert watts to VA, multiply by 1.4.

4. Total the VA of the number of pieces of equipment.

5. Multiply the total VA by 1.2 to allow for future expansion.

6. Make sure that the total VA requirement of supported equipment does not exceed the UPS VA rating of the equipment you buy.

Disaster Recovery

If you can't afford to shut down your business when disaster hits, you should also have a comprehensive disaster recovery plan. Anticipate what could go wrong, and specify policy, procedures and technology that will minimize your losses when disaster occurs.

Disaster recovery options include network diversion, PBX bypass to a standby PBX or key system, wireless or cellular services, pagers, alternate answering points and help desks.

Steps to take when designing a disaster recovery plan:

1. Decide upon the level of service that will be maintained during a disaster.

2. Identify redundant paths in the event of a cable cut.

3. Identify critical areas, requiring complete communications, areas requiring partial communications and those that will only require minimum communications.

4. Determine where calls will be diverted for incoming ISDN, digital and analog lines. Ensure that there are policies and procedures in place to process the incoming calls.

5. Decide how the organization will communicate internally in the event of an outage.

6. Identify where help desks or command centers will be located during an outage. Make sure there are sufficient resources available to accommodate the additional load.

7. Have regular drills and tests. This will ensure that power protection, fallback communications and other recovery systems will work in real-world emergencies, and that employees know what to do when the lights go out.

The most advanced power protection systems now offer integrated power management software. This software typically tracks power conditions, records load levels, tracks battery usage and temperature and can initiate orderly shut-down procedures if the battery is running low and the procedure is not initiated by a human. Some power management software can even forecast the power needs of a network.

Examine your company's business processes to determine which systems are critical to the actual running of the business. Back up those systems with the power protection that best fits your company's price/performance requirements.

Provisioning and Transport Services

Provisioning Your Phone System, Dialing and Local Service

Provisioning is the specification, ordering and installation of trunks, lines and services into and out of your telephone system so that you can make and receive calls to other people and companies.

How To Determine How Many Phone Lines You Need

Determining the number and type of outside lines coming into your telephone system is an inexact science. Overkill is marginally better than not enough lines-too few lines and your incoming callers will get busy signals and your employees won't be able to get an outside line. Then again, you don't want so many trunks that you're paying for capacity being unused.

The easiest way to estimate the number of trunks you need for your phone system is based on the number of people using your system. As the number of people on the system increases, the proportion changes. Efficiency of use increases as the number of trunks increases. For a 50 person office, 30 trunks is plenty. If you have 100 people and average telephone use, (no high-traffic call centers), 25 combination (also called bothway trunks; they handle either incoming or outgoing calls) trunks will probably be more than enough. One hundred trunks will be too many for 400 people.

As the number of users decreases, you will need more trunks as a percentage of users. The smallest systems might use one trunk per person.

If you are using DID trunks, which are incoming only, with combo trunks for outgoing, you will need more trunks altogether because unused DID trunks can't be used for outbound calls when all of the bothway trunks are busy.

A standard T-1 circuit's individual 24 channels can be set as either DID or bothway trunks.

PRI (ISDN) T-1 circuits are more flexible because any of the 23 PRI channels can carry either incoming direct-dialed or outgoing calls.

Find out what kind of outside lines your prospective phone system supports. If you need more than about 12 trunks, it's usually cheaper to go with a digital T-1 line (the equivalent of 24 analog lines) than a number of analog DID or CO/bothway trunks. Compare the cost of the interface for your telephone system plus the ongoing cost/capacity of the two types of service.

Traffic Engineering

There are software programs available based on traffic engineering principles that help estimate the number of trunks you'll need for your phone system. You can also use special statistical tables and coefficients to figure out how many trunks you should have.

Traffic engineering is a very complicated (and sometimes not very accurate) way of computing the number of trunks needed for a given telephone system. Traffic engineering uses statistical probabilities to determine the number of trunks needed for a phone system based on assumptions you make about the service level you'll provide to your users.

Traffic engineering equations require you to know what your busy hour is — the number and length of incoming and outgoing calls your company handles in the busiest hour of the busiest day at the busiest time of the year. You need to collect a lot of data to figure out when the busy hour is, and then you need to measure the amount of usage your trunks are getting during this busy hour.

You can get inbound and outbound call volume and duration data from your PBX, and call accounting systems will generate call data reports. Phone bill analysis is another, not-so-accurate way of collecting call data, because unless all of your incoming calls are over a toll-free number or digital circuit that's Caller ID-capable, you won't be able to count incoming as well as outgoing calls and your data will be faulty.

Traffic engineering is generally used by telephone companies (ILEC or CLEC) and by analysts for big private companies.

Unfortunately, accurate data about the calling patterns can be difficult to gather, and the engineering estimates can't be trusted if they are not based on accurate data. Secondarily, and no less important, the call data used by traffic engineering calculations is historical information, so the calculations don't include a growth factor.

Carriers and Providers

For years you've been able to choose your long distance company and, depending on where you live, you've also been able to choose your regional (local toll or one-plus) telephone service provider. Some of us can select our local telephone companies as well, because the Telecommunications Act of 1996 opened up the telephone industry to wider competition.

Greater competition and choice is leading to a variety of services, from well-recognized companies and from start-ups. Cable television companies, utility companies, CLEC (Competitive Local

Exchange Carriers) and Internet service providers (ISPs) may compete with telephone companies to provide you with telephone and other communications services.

Hundreds of companies now provide interstate and international long distance telephone service, and the Telecommunications Act of 1996 introduced competition into the local service market. Some companies offer service throughout the United States, and others offer service only in certain parts of the country. Some design their service to appeal to certain types of callers, others design their service to appeal to people who support a specific charity by donating a percentage of the caller's phone bill to that cause. Most companies compete on the basis of price.

Existing providers are not losing money or customer willingly; they're offering dial-around services and special discount plans. You can probably save money, right now, without switching companies by asking about and signing up for savings plans offered by your current long distance company.

Local Service

Local trunks are a network of circuits running between local central offices and between local and toll switching systems.

Local calling was, until recently, provided solely by the regulated monopoly, Incumbent Local Exchange Carrier (ILEC) such as a local Regional Bell Operating Company (RBOC), GTE or other independent company. Local calling was not subject to competitive marketing and was heavily regulated at the state and federal level. The Telecommunications Act of 1996 lifted these restrictions and now local calling is open to competitive providers.

A few CO codes have been reserved on a network-wide basis for special uses:

555	Toll (intralata) Directory Assistance (DA)
950	Feature Group B Access (FGB)
976	DIAL-IT Services
011	International Direct Distance Dialing (IDDD)
411 (or 1411)	Directory Assistance (DA)
611	Repair Service
911	Emergency Service

Local calls are usually dialed using the seven digit CO code plus the station number. In areas where there has been a new code added using an NPA (Numbering Plan Area = area code) overlay, you have to dial ten digits for local calls.

Toll calls, both intra and inter lata, are usually dialed as 1+NPA+7 digits. Some areas, like New England and Chicago, still allow intralata local toll calling using seven digits without a 1 prefix.

How you are charged for local calling depends upon how you order your customer interface (CI) trunks, and where you're ordering. You may be charged for local calling in one of several ways:

Charges for Local Calls
Flat Rate You pay a single monthly charge and are allowed to make as many local calls as you can without additional charges.

Message Rate You're billed a lower monthly charge than the flat rate, and are allowed a limited number of calls (usually under 30) per month without additional charges. Additional calls are then billed on a per call basis.

Measured Rate This is a very small monthly service charge, but you pay for every local call based on the quantity and duration of the call.

LATAs (Local Access Transport Areas)

Local Access Transport Areas (LATAs) are geographic areas in which

231

there is a 'community of interest'. A LATA consists of more than one Local Calling Area (LCA) with toll charges for calling between LCAs.

A LATA may serve more than one area code or one area code may have more than one LATA. LATAs can cross state boundaries where there is a community of interest. States with small populations might have only one LATA.

LATAs were mandated by the Justice Department with the divestiture of AT&T in 1984 as a means to identify the traffic that could be carried by ILECs (Incumbent Local Exchange Carriers) versus the type of traffic that had open competition and could be carried by Interexchange Carriers (IXCs).

Calls between Local Calling Areas within the same LATA were deemed to belong to the ILEC serving that LATA and were not subject to competition. With the 1984 equal access ruling, intralata toll calls were automatically carried by the serving LEC unless the caller directs the calls to an IXC by dialing a Carrier Access Code (CAC) before the called number.

Equal Access has been superceded by The Telecom Act of 1996, which opened up the intralata market to competition.

Local Number Portability (LNP)

The 1996 Telecommunications Act requires all local exchange carriers (LECs) to provide, to the extent technically feasible, number portability.

See The Telecommunications Act of 1996, the Local Loop and CLEC (Competitive Local Exchange Carriers) chapter for more information about the Act.

LNP lets you change local service providers without changing telephone numbers. The LEC that loses the customer must delete the end user customer information from the LNP database. The new LEC must add the customer and show the new routing code (Location Routing Number or LRN). The changes should be coordinated between telephone companies and be transparent to the end users, i.e., no unusual dialing delays, etc.

Rather than selecting a particular architecture, the FCC has established performance criteria that an ILEC's long-term number portability architecture must meet.

Here's the FCC's verbiage, condensed:

Any long-term method must support existing network services, features, and capabilities, efficiently use numbering resources, not require end users to change their phone numbers, not require telecommunications carriers to rely on databases, other network facilities, or services provided by other telecommunications carriers in order to route calls to the proper termination point, not result in unreasonable degradation in service quality or network reliability when implemented, not result in any degradation of service quality or network reliability when customers switch carriers, not result in a carrier having a proprietary interest, be able to accommodate location and service portability in the future, and have no significant adverse impact outside the areas where number portability is deployed.

Location Routing Number (LRN), as proposed, is the only specific method currently supported by the industry that complies with all these criteria.

The FCC established timetables for the deployment of LNP by dividing the country into two segments-the one hundred largest Metropolitan Statistical Areas (MSAs) and the balance of the country. All LECs within the 100 largest MSAs had to provide LNP by December 31, 1998. After that, all other LECs-those outside the 100 largest MSAs-must be prepared to provide number portability within six months of the date service is requested by another carrier.

The incumbent local exchange carriers (ILECs) are responsible for routing calls to CLEC customers in adjacent areas. If a call is sent via an interexchange carrier (IXC), it is the IXC's responsibility to deliver it to the proper ILEC. If a call is between two ILECs, the originating ILEC is responsible for determining the proper routing.

LNP requires that the telephone company be Signaling System 7 (SS7) capable, and, the SS7 software version used must be Advanced Intelligent Network (AIN) capable in order to be able to query LNP databases.

Cellular and broadband PCS carriers must have been able to query a number portability database, or have made arrangements with other carriers to do these queries, by December 31, 1998, and were to implement long-term number portability by June 30, 1999.

> **See the Data Communications and Digital Voice chapter for more about Signaling System 7.**

Long-term number portability requires a national system of regional databases managed by an independent third-party local number portability administrator(s) (LNPAs) selected by the North American Numbering Council (NANC). Selection of LNPAs in certain areas has been slow, and legal holdups and technical unfeasibility are also contributing to LNP implementation delays.

Local exchange carriers are providing interim number portability methods, such as Remote Call Forwarding (RCF) or Direct Inward Dialing (DID), in accordance with the Order.
The FCC has allowed carriers to determine where in their networks it is most efficient to query the LNP database. The terminating carrier can't perform LNP database queries unless all of the involved carriers agree, or unless they are the only carriers equipped to perform a query.

Directory Assistance

Directory Assistance (DA) is provisioned differently in different areas depending on state tariffs. In most cases, local DA is provided by dialing 411 or 1+411.

Some state tariffs require the calls to be dialed as 1+ if there is a charge for the service. A few states allow 411 without the 1 prefix even though it is a chargeable call. A few states don't allow DA to be dialed as 411 or 1+411 at all. Because it is a chargeable call, these states require that it be dialed as 1+555-1212 or 1+NPA+555-1212. New Mexico also provides DA in Spanish, dialed 1+555-1313.

All Directory Assistance calls route to a special local switching center where the calls are answered by live operators. Interlata DA is always dialed 1+NPA+555-1212 and routes to the trunk's assigned IXC switch.

Service Access Codes (SAC)

Service Access Codes (sometimes referred to as Special Access Codes) are special NPAs (area codes) that have NXXs (the next three digits after the area code) that are dedicated to specific carriers. The most common SAC in use today is 900, for pay-per-call services. There's also a 500 service that provides "follow me" services. Charges for 900 and 500 service are usually but not always billed with your local telephone bill.

700 Service

Code 700 is a special code that is reserved for IXCs (Interexchange Carriers) to use however they like. Some IXCs use 700 to bypass the ILEC (Incumbent Local Exchange Carrier) in routing local and intralata toll calls .

Almost all IXCs use 1+700+555-4141 as a circuit identification number. When dialed, this number goes to the carrier identified by the line PIC (Primary Interexchange Carrier) or dialed CAC (Carrier Access Code, as in 1010CAC) of the line you are calling from. A recording tells you the circuit number of the trunk you're using to dial out.

Feature Group B

Feature Group B (FGB) was pre-equal access technology. After divestiture, this method was developed quickly to enable routing of LD (Long Distance) calls to other carriers besides AT&T. This interim solution provided a standard numbering strategy for network access.

When a Feature Group B call connects to an IXC, the IXC sends a tone back to the caller, and the caller enters a calling card or credit card number and the telephone number being called. Billing information is confirmed, the IXC processes the call and then delivers it to its destination. Feature Group B access is still used frequently for calling card services, but has otherwise been replaced with equal access ("dial one") dialing.

Interstate Access Charge System

Income from local telephone service has never been high enough to cover local telephone company network operating and maintenance costs. So, long distance companies have been charged to make up the difference.

IXCs use the local networks to complete their long distance calls. The charges IXCs pay to local carriers (ILECs) are called "access charges." Each long distance telephone call you make includes per-minute fees that your long distance carrier pays to both your local telephone company and to the local telephone company of the person you call.

In 1997, the FCC reformed its system of interstate charges to make the system compatible with the development of competition for all telecommunications services, including local telephone service.

Federal Subscriber Line Charge

Local phone companies recover some of their local operating costs through a monthly "subscriber line charge" that appears on the local telephone bill of businesses and some residential customers. Sometimes the charge is called the "federal subscriber line charge" because it is regulated by the FCC and not by state public utilities commissions.

As part of its access charge reform effort, the FCC reduced subsidies that keep the subscriber line charge low for residential consumers beyond their first telephone line. The second and any additional telephone lines are called "non-primary" lines.

The access reform plan reduces subsidies for non-primary residential telephone lines and shifts the method by which local telephone companies recover the costs of providing local loops. This is part of an overall plan to substantially reduce per-minute long distance phone rates. Many consumers with more than one residential telephone line will be better off under the new system — especially those who make a large number of long distance calls.

"Local Loop" refers to the outside telephone wires, underground conduit, telephone poles, and other facilities that link each telephone customer to the telephone network. For more about Local Loops, see the chapter entitled The Telecommunications Act of 1996, the Local Loop and CLEC (Competitive Local Exchange Carriers).

The system of subsidies that pay the balance of local loop costs has forced everyone to pay higher long distance rates. The subsidies also have stifled local competition because new companies must compete against subsidized rates charged by existing local phone companies.

TRUNKS, DIALING PLANS AND LOCAL SERVICE / CHAPTER 25

The FCC is allowing local telephone companies to raise the flat fee on non-primary residential telephone lines so that those lines are no longer subsidized, or receive less subsidy. The increase in price for non-primary telephone lines is supposed to make customers pay for the cost of their local facilities.

> **New technologies have developed to take advantage of loopholes in federal access charges that long distance providers pay to local providers, and vice versa. See the Voice Over the Internet Protocol (VoIP) chapter for more.**

T-1 (Transmission One) Carriers

The T-carrier system, introduced by the Bell System in the 1960s, was the first system that supported digitized voice transmission. It was developed by the telcos to allow more voice calls to be transported over copper wires between their local offices. T-1 rapidly became the backbone of long distance toll service and is the primary transport method for local transmission between central offices.

T-1 is full-duplex, using two pair (four) wires; one pair to receive and one pair to send. Dedicated copper T-1 circuits terminate using a RJ-48C connector.

The term "T-1" originated as a description of a very specific type of physical equipment: selected cable pairs and digital regenerators at 6,000-foot intervals. Later developments improved on the capacity of the original T-1 repeaters. Placing the regenerators at 3,000-foot intervals doubled the bit rate to 3.152 Mbps. (called DS-1C).

Copper-based T-1 circuits are billed based on the additional cost of repeaters; the longer the circuit, the higher the price. Despite the largely fiber optic physical plant installed today, the mileage-sensitive tariff structure remains.

T-1 (Transmission One) signaling reduces analog voice traffic to a series of binary coordinates that represent sound.

A single voice signal in digital form takes 64,000 bps. This is known as digital signal level zero, or DS-0. When T-1 originated, 1.5 Mbps was about the highest rate that could be supported reliably over the one mile distances between manholes in large cities. (Manholes were used for locating cable splices, and were sites for signal boosters or regenerators.)

Each T-1 channel runs at 64 Kbps. A T-1 circuit is a 23 (ISDN PRI) or 24 channel (Standard T-1) digital transmission link with a capacity of 1.544 Mbps (1,544,000 bits per second).

The T-carrier system uses pulse code modulation (PCM) to translate analog voice into a stream of digital bits.

Pulse Code Modulation (PCM)

Pulse Code Modulation (PCM) is the existing worldwide standard for digital voice. PCM conversion between analog and digital happens using a CODEC (COder/DECoder). CODECs convert analog voice into a digital bit stream in two steps:

1. Pulse Amplitude Modulation (PAM)
The incoming analog signal representing a continuous varying voice signal is measured 8,000 times per second. The modulator uses the resulting sample to send a very narrow squared-wave pulse whose voltage (height) is the same as the analog signal's to the coder.

2. Digital encoding using Time Division Multiplexing (TDM)
The height of the pulse is then converted to a digital value by a coder, an analog to digital converter. The output is an 8-bit code word (hence "pulse code") representing the voltage of the

pulse, which was the value of the analog input when the sample was taken.

The two-step PCM process converts an analog voice signal to a digital stream of 64,000 bits per second (64 Kbps), which is 8 bits x 8,000 samples/sec.

The digital representation of a signal can take on only a small number of values, at discrete steps. An analog signal, by contrast, has almost infinite variability. Therefore, at the precise time of a sampling, an analog input is seldom exactly the same as a possible digital output step. The coder, however, must make a selection, and will pick the closest digital value.

The rate of 8,000 samples per second comes from the Nyquist theorem. This theorem shows that an analog reconstruction from digital data can contain all the information of the original analog signal, if the sampling rate is faster than two times the highest frequency found in the original signal.

If sampling is not rapid enough, the resulting digitized points can represent more than one analog signal. This phenomenon is called aliasing and produces unintelligible sounds. Aliasing is avoided by filtering the input signals to eliminate high frequencies.

T-1s transmit at frequencies higher than 300 Hz and lower than 3,300 Hz. Filtered voice signals can't be higher than half the sampling rate of 8,000 times per second, or 4,000 Hz (4 kHz), and the usable upper limit is 3,300 Hz. Filtering out the low end, to block 60 Hz hum, puts the practical lower limit at 300 Hz.

Since it uses digital regeneration, TDM offers relative immunity to noise. Digital signals like TDM also lend themselves to switching, control, and maintenance without being demultiplexed into analog signals as a part of the communications process.

Standard T-1

A standard T-1 has twenty-four 64 Kbps voice channels plus an 8 Kbps channel for control and synchronization and runs over two pair of standard copper wires. The standard T-1 rate is the same in North America, Japan and Australia.

At 64 Kbps each, the aggregate data rate of 24 channels is 1,536,000 bps. The 8 Kbps control channel brings the total bandwidth to 1.544 million bps (Mbps).

A standard T-1 circuit's individual 24 channels can be set as either DID or bothway trunks. PRI (ISDN) T-1 circuits are more flexible because any of the 23 PRI channels can carry either incoming direct-dialed or outgoing calls. (An ISDN PRI-based T-1 is 23 voice/data channels and one 64 Kbps signaling (or D) channel.)

See the ISDN chapter in this section for more about ISDN PRI service.

In European and other CCITT countries, 30 voice channels and two signaling channels are multiplexed at a rate of 2.048 Mbps (2,048,000 bits per second), and called data signal level one, or DS-1.

Channel Service Unit (CSU) and Digital Service Unit (DSU)

The piece of equipment connected to a T-1 circuit at the customer end is generally a CSU (Channel or Customer Service Unit) or DSU (Digital Service Unit). Before divestiture, the CSU/DSU was considered part of the public network and was owned by the phone company.

The term CSU used to mean a channel bank, which is an electromechanical de-multiplexer that connects incoming T-1 circuits to analog PBX interfaces. CSUs proceeded DSUs, which are digital interfaces from the network straight through to the CPE. Today these terms are used more or less interchangeably, and the "C" in CSU means Customer.

A CSU/DSU contains the last signal regenerator on the T-1 circuit before it terminates at the end user's equipment (DTE, Data Terminal Equipment). It also has a mechanism to put the circuit into loopback for testing from the central office. A third function of the CSU/DSU is to monitor the T-1 data stream for signal irregularities or loss and to generate a "keep alive" signal when the attached DTE fails to deliver data or is disconnected.

To regain the ability to set up a loop-back for testing after CSU/DSU units became the property of their customers, carriers started installing a device to mark the demarcation point between the local loop and the customer premises wiring. Called a "smart jack," the interface generally goes in the wiring MDF (Main Distribution Frame) room of a large building, or outside of smaller buildings. The FCC accepts that the smart jack is not CPE (Customer Premises Equipment), so it can be owned and supplied by the carrier.

The transfer of CSU/DSU ownership from the Central Office to the customer means that the CO is no longer responsible for powering the T-1 circuit.

When the CSU/DSU belonged to the network, the carrier supplied DC power through the signal leads. Send and receive data pairs were balanced and isolated from ground by transformers, so each pair could be used as one power lead. AT&T originally kept the CSU/DSU functioning during a local power failure at the customer site to avoid central maintenance alarms.

Now that they no longer are responsible for the CSU/DSU, carriers generally refuse to or can't (optical fibers have no way to provide power) power T-1 circuits. The FCC now allows carriers the option of powering the CSU or not, for both copper and fiber-based circuits. Policy has come down to no power provided for new T-1 installations. This means that your T-1 circuit should be backed up, either with power failure equipment or by standard CO analog trunks.

How to Buy a T-1 Circuit

Dedicated service has a recurring monthly charge know as a Loop or POP (Point of Presence, the local office of the LD carrier) charge. This charge is based on the mileage to the carrier's POP from your CPE.

The one-time installation charge for the T-1 circuit varies, again based on the mileage and the installer. In many cases you can reduce the installation charge, or get it waived, especially in populous states. Where there is heavy competition for long distance minutes, local and long distance carriers have a harder time selling T-1s because long distance direct-dial calls are cheaper.

A lengthy T-1 can involve three carriers: a local operating telephone company at each end and a long-distance company in between. Each local phone company provides access to the switching offices of the long distance provider.

T-1 circuits in the US are available from local and long distance providers, although usually the local telco provides the local copper wire from your location to their nearest central office.

Outside the United States, the local PTT (the national Postal, Telephone, and Telegraph authority) provides the entire circuit within its serving country. PTTs interconnect their circuits at national borders for multinational networks.

How To Decide If You Should Buy A T-1 Circuit

How Many Trunks Do You Need?
Find out what kinds of circuits your prospective phone system

supports. If you need more than about 12 trunks, it's usually cheaper access-wise to go with a digital T-1 line than a number of analog DID or CO trunks. Include the cost of the T-1 interface for your telephone system when you compare the ongoing costs of T-1 circuits and analog trunks.

Do You Need Caller Line ID Or Dialed-Number Identification?
If your IVR (Interactive Voice Response) system will be answering toll-free or 900-based calls and it must respond to ANI (Automatic Number Identification —the number of the caller) or DNIS (Dialed Number Identification Service - the number the caller dialed) then go with T-1.

You can get similar ID services over analog circuits (Caller ID), but the cost per line is higher. DID interface cards in the phone system cost extra. Additionally, Caller ID can be blocked by the caller; ANI/DNIS info can't be blocked at all on 800/900.

How Much Does Installation Cost and How Soon Do You Need Service?
Installing a T-1 is more expensive than installing 24 analog lines; usually about 20% more. Installation lead time for a T-1 is about three weeks; analog lines can be working in your office in a few days.

What Do The Alternatives Cost?
Compare the relative costs of CO and DID trunks and T-1 circuits for your particular location. In some places T-1s cost nearly twice as much as 24 analog trunks per month, in others T-1s are cheaper.

How Much Will It Cost To Upgrade Your Phone System?
PBXs don't automatically come T-1 capable; they need an interface. This card can be quite expensive. ($2,000 plus). If you have IVR (Interactive Voice Response), make sure that ANI and DNIS information will pass through the PBX to your IVR system so that calls can be handled smarter by your automatic answering system.

What are Your Performance And Reliability Requirements?
Digital trunks are more reliable. There is less noise, and you get faster connections.

The downside is, if one analog line goes down you're not out of service, but if your single T-1 connection fails, all of your trunks go dead. There's a consolation: T-1 repairs often happen faster (they can do a lot of troubleshooting remotely and they try harder because you're paying more) than repairs to analog trunks.

It's a good idea to have backup for your T-1, either analog trunks or power failure equipment. Factor these costs into your buying equation.

Long Distance Services

The North American Numbering Plan (NANP), International Calling, The Presubscribed Interexchange Carrier (PIC), Charge CIC (Carrier Identification Codes) and Dial-Around, Toll-Free Services, and How to Buy Long Distance Services

Long distance systems are a network of switching systems and circuits that haul telephone traffic between local exchange areas (Local carriers are called LEC for Local Exchange Carriers or, if they used to be Bell-related, ILEC for Incumbent Local Exchange Carrier.). Long distance carriers are usually called IXC or Interexchange Carriers.

Your long distance calls leave your business over LEC facilities, are transferred to your IXC's network, can be transferred to and from a competing carrier, and are finally sent back to the LEC at the destination end.

The North American Numbering Plan (NANP)

The NANP (North American Numbering Plan) was designed as a closed ten digit national numbering plan. The routing codes for the NANP have three parts: a three digit area code or Numbering Plan

Area (NPA) code, a three digit Central Office (CO) code (or NNX), and a four digit station number.

Together, these ten digits comprise the network "address" or "destination code" for each telephone number.

International Calling

International dialing uses an open numbering plan capable of handling from 7 to 15 digits.

IDDD (International Direct Distance Dialing) was first introduced in 1970. The prefixes 011 for station to station calls and 01 for operator assisted calls are used for IDDD calling according to the NANP (North American Numbering Plan).

The worldwide numbering plan divides the world into nine world zones. The USA is part of world zone 1 that also includes Hawaii, Alaska Canada, Puerto Rico, the US Virgin Islands, Bermuda and other Caribbean Islands.

Mexico, geographically part of North America, chose to join the Latin American countries of Central and South America in world zone 5 and is dialed using the international format.

International telephone numbers are dialed with a Country Code (CC) followed by the country's National Number (NN). International agreements include a restriction that the NN will not exceed 15 digits.

Country codes may have one, two or three digits. Each country's world zone number is the first digit of its CC.

The United States has been assigned the one digit country code 1 (and is world zone one), Belgium's CC is 32 and Portugal has three digit country code, 351. Europe is divided into so many countries needing two-digit country codes that the area is assigned two world zones: 3 and 4.

The world numbering zones are:

1 North America (NANP)
2 Africa
3 Europe
4 Europe
5 Central America, South America, Mexico and some Caribbean points
6 South Pacific (Australia, New Zealand and many island groups)
7 Russia (and former USSR countries)
8 North Pacific (Eastern Asia)
9 Far East and Middle East

There's a list of country codes at the end of this chapter.

International calls are transported by International Carriers (INCs). Most INCs in the USA are IXCs, carrying international and domestic interlata traffic both.

When you make a long distance call, the IXC uses your local phone company's facilities (switching equipment, connections, and wires) to pick up the call from you, and the local phone company at the other end's facilities to complete your call. The long distance company is required to pay the local companies for the "access" they provide to get to you.

The Presubscribed Interexchange Carrier (PIC) Charge is an access charge that long distance companies pay to local telephone companies for each residential and business telephone line presubscribed to their service. If you don't presubscribe to a long distance carrier, the local telephone company may bill you for the PIC Charge anyway.

Under the old rules, IXCs paid a per minute fee to the local telephone companies for access. As of January 1, 1998, IXCs pay a flat rate, per-telephone line charge plus a lower per-minute charge for local access.

The PIC Charge is designed to recover the local telephone companies' interstate local loop costs that are not recovered through subscriber line charges. The rationale behind the change in access charges is that because the costs of the local loop do not depend on usage, a flat-rated charge better reflects the local telephone company's costs of providing service. The Presubscribed Interexchange Carrier Charge rates vary from state to state based on the actual cost of providing local phone service in each area.

Because the long distance market is competitive, the FCC does not directly regulate long distance company charges. As a result, long distance companies take different approaches as to how they pass on PIC charges to their customers. Increases in PIC charges paid by the long distance companies theoretically have been offset by reductions in the per-minute access charges paid, so if your carrier's monthly charges go up, your per-minute pricing should go down.

CIC (Carrier Identification Codes) and Dial-Around

In 1982 AT&T agreed to divest itself of the regulated local service portion of its business by January 1, 1984. The local service companies retained the Bell logo and the entities were labeled Regional Bell Operating Companies or RBOCs.

The Carrier Identification Codes are three or four digit codes that identify a long distance carrier, or IXC (Interexchange Carrier). Carrier identification codes (CICs) tell the local phone company (ILEC or Incumbent Local Exchange Carrier) which long distance company you want to use to make a toll call. ILECs use CICs to route traffic and to bill for interstate access service provided.

CICs let you use any telecommunications service provider by dialing a carrier access code, or CAC. Since long distance de-regulation, CICs have been unique three-digit codes (XXX) and CACs were five-digit codes incorporating the CIC (10XXX).

AT&T reserved PIC 288 for itself because on the telephone dial, 288 is ATT. MCI uses PIC 222 and Sprint uses 333. The MFS PIC is 440.

Due to a shortage of CIC codes (three digits allow only 1000), CIC numbers were expanded to four digits in 1993, allowing up to 10,000 IXCs. Three digit CICs became four digit CICs with a leading zero.

The move to four digit CIC codes made approximately 10,000 numbers available for assignment. More numbers lets more companies compete, and more competition means lower prices and better service for consumers.

We've all seen the low, low prices advertised everywhere for dial-around services. (1010321, 10109000 and 1010345 are three that have been advertising heavily.) Each dial-around service is priced differently, with different combinations of fixed cost, minimum minutes and per-minute costs. Be sure you understand the actual price of a service before you inadvertently spend more than you intend.

Calls made using long distance dial-around may be billed directly by the long distance company that you used, by an agent billing for that long distance company, or through your local telephone company.

Toll-Free Services

Toll-free numbers are used by businesses to strengthen relationships with customers. But busy signals, long queues and unanswered lines can damage both the concept of this service and good consumer relations.

Your company pays for toll-free numbers so that you are easily accessible to current and potential consumers. You want to make it easy for callers to place orders, check their orders and ask questions. If you use a toll-free service but fail to deal with your customers smoothly and efficiently, your customer satisfaction will suffer.

Research shows that 5% of the calls into toll-free numbers receive a busy signal. An additional 7% of the calls are abandoned when customers are left on hold too long or encounter a lengthy automated IVR (Interactive Voice Response) menu.

The length of time people will wait on hold usually depends on the reason they are calling. Customers are fairly tolerant of wait time, as long as they get through and get their problem resolved. Acceptable wait time is about a minute.

When a toll-free call is dialed, the local central office (CO) sees the "1", recognizes the call as long distance, then ships the call to a larger CO. That switch sees the 800, 888 or 877, looks at the next six digits, and performs a database query to decide which long distance carrier will handle the call, and where to send it.

Database sophistication gives you have many routing choices for each toll-free number. Routing can change by day or by time of day, at a traffic-volume threshold, by the location the call originated, or based on a mathematical formula.

It frequently is more cost effective to use an alternative carrier for toll-free service. 800 portability lets you can keep your number when you change carriers.

Toll-free directory assistance for 800, 888, and 877 numbers is available by dialing 1-800-555-1212.

Toll-free numbers are currently reserved on a first come, first served basis. The 877 numbers were first offered for general reservation on April 5, 1998. The placing of an order for an 877 number prior to the general release of 877 numbers did not guarantee that anyone would actually receive that number.

Entities called Responsible Organizations (RespOrgs): toll-free service providers and carriers, have access to a database that contains information regarding the status of all toll-free numbers.

The industry has plans to introduce 866, 855, and other codes for toll-free calling once 877 numbers are assigned.

You can follow the ins and outs of the FCC at http://www.fcc.gov. The Common Carrier Bureau, which regulates interstate telephone services, has a page on this site that gives up-to-date information on toll-free numbers.

It is no longer legal to buy and sell vanity toll-free numbers. In the Second Report and Order released on April 11, 1997, the Commission concluded that the practices of hoarding and brokering toll free numbers are not in the public interest and that parties that hoard and broker numbers will be subject to penalties.

How to Buy Long Distance Services

Call a long distance company directly and ask them if they provide service in your area, or ask your local telephone company for a list of long distance companies providing service in your area. You can check the yellow pages under telephone companies or search for long distance providers on the Internet.

If you don't select a primary long distance company when you order a new line, your local telephone company may select a company for you.

Long distance calls should be billed in six second, rather than one minute increments, whenever possible. You can save a substantial amount of money, especially if you make many short calls, by insisting on six second billing.

Long distance companies have a variety of calling plans and rates. For the best comparison, provide specific information about your calling patterns, like the geographic areas that you call, the time of day that you place calls, the average length of your conversations, and any special services that you want a long distance company to provide. Check to see if you can get local service from your long distance provider. Combining local and long distance minutes through the same company might lower your per-minute rates.

Your local telephone company will charge you $5 to switch to another long distance company. Your new long distance company may agree to reimburse you for this fee, but is not required to do so.

When you're collecting bids on long distance services, print out your trunk report if you have a call accounting system). This will give prospective bidders your approximate call volume and the geographic areas you call most. Ask for incremental pricing based on your call volume, and ask for discounts for signing 2, 3, 4, or 5 year contracts. Be sure to ask each company about its per-minute rates and special calling plans. You should make sure that you are getting the best deal for the types of calls you place. Don't forget to incorporate your expected growth.

Companies compete for your telephone business. Use your buying power wisely and shop around. Long distance companies are taking very different approaches to whether and how they are changing charges to their customers to reflect the PIC charges they pay.

Unless you are a large company with specific needs (frame relay, T-1's) there isn't much difference between long distance carriers except price, the appearance of the invoice and the number of commercials the carrier runs.

Periodically conduct long distance audits. It's a good idea to check your usage vs. capacity, and check competitive rates at least once per year.

Verify intrastate rates, which vary widely by state, and are frequently higher than interstate rates.

International Country Codes

These codes are subject to change, so check with your long distance company for current information and dialing instructions.

Country	Code	Country	Code
Albania	355	Costa Rica	506
Algeria	213	Croatia	385
American Samoa	684	Cuba	53
Angola	244	Cyprus	357
Argentina	54	Czech Republic	42
Armenia	374	Denmark	45
Aruba	297	Dominican Republic	809
Australia	61	Ecuador	593
Austria	43	Egypt	20
Azerbaijan	994	El Salvador	503
Bahamas	809	Ethiopia	251
Bahrain	973	Fiji	679
Bangladesh	880	Finland	358
Barbados	809	France	33
Belguim	32	French Antilles	596
Belize	501	French Guiana	594
Bermuda	441	French Polynesia	689
Bolivia	591	Georgia	995
Bosnia-Herzegovina	387	Germany	49
Botswana	267	Ghana	233
Brazil	55	Gibraltar	350
Bulgaria	359	Greece	30
Cambodia	855	Greenland	299
Canada	1	Guadeloupe	590
Caribbean Islands	809	Guam	671
Central African Republic	236	Guantanamo Bay (Cuba)	5399
Chile	56	Guatemala	502
China, PRC	86	Guinea	224
Colombia	57	Gyuana	592
Congo	242	Haiti	509
		Honduras	504

Hong Kong	852	Netherlands	31
Hungary	36	Netherlands Antilles	599
Iceland	354	New Zealand	64
India	91	Nicaragua	505
Indonesia	62	Nigeria	234
Iran	98	North Korea	850
Iraq	964	Norway	47
Ireland	353	Pakistan	92
Israel	972	Panama	507
Italy	39	Paraguay	595
Ivory Coast	225	Peru	51
Japan	81	Philippines	63
Jordan	962	Poland	48
Kenya	254	Portugal	351
Korea	82	Romania	40
Kuwait	965	Russian Federation	7
Laos	856	Rwanda	250
Latvia	371	Saipan	670
Lebanon	961	Saudi Arabia	966
Liberia	231	Serbia	381
Libya	218	Sierra Leone	232
Liechtenstein	4175	Singapore	65
Lithuania	370	Somalia	252
Luxembourg	352	South Africa	27
Macedonia	389	Spain	34
Madagascar	261	Sri Lanka	94
Malaysia	60	Sudan	249
Malta	356	Suriname	597
Martinique	596	Swaziland	268
Mexico	52	Sweden	46
Monaco	377	Switzerland	41
Mongolia	976	Syria	963
Montenegro	381	Taiwan	866
Morocco	21	Tanzania	255
Mozambique	258	Thailand	66
Nepal	977	Tunisia	216

Turkey	90	Vietnam	84
Uganda	256	Western Samoa	685
Ukraine	380	Yemen Arab Republic	967
United Arab Emirates	971	Yugoslavia	381
United Kingdom	44		
United States	1	Zaire	243
Uruguay	598	Zambia	260
Venezuela	58	Zimbabwe	263

New Area Codes

Split vs Overlay

Before 1995, the middle digit of an area code was either one or zero. That's no longer the case. The need for more numbers necessitated allowing other numbers as middle digits .

The increase in the number of ways to reach people (like fax, cellular, PCS, pagers and the Internet) is responsible for the sudden growth in the demand for phone numbers. As existing area codes become saturated with phone numbers, more have to be added.

New area codes mean that security systems, programmable phones, modems and all devices that have automatic dialing mechanisms need to be reprogrammed, including your PBX or key system.

The area code undergoing the world's greatest change is 809, which had served as a catch-all way to reach tropical destinations like Puerto Rico and the Virgin Islands. Now the territory is being split into 17 new area codes so that each island will have its own area code.

Some US territories, like Guam and the Commonwealth of the Northern Mariana Islands, will get area codes for the first time. Their international country dialing codes became their area codes.

Area Code Implementation

There are two ways a new area code can be introduced. The best and least disruptive method is to overlay it over the existing area code. A more expensive and inconvenient implementation splits the geographic area that the existing are code serves and assigns part of it to a new area code.

Overlays mean all local callers have to dial 10 digits - the area code of the person called, plus the regular number. Overlays also saves customers from having to pay for area code changes to their stationery, business cards, work trucks and websites.

Overlay implementation plans install the new area code over an existing area code. Both area codes share the same geographic boundaries and the same geographic identity. New customers receive the new area code. Existing customers keep their existing area code and telephone number.

Overlays are less disruptive for current customers than a geographical split, the other method used to implement new area codes.

Geographic splits divide the existing area code into two separate geographic sections. The existing area code serves one geographic area and a new area code is assigned to the other. Telephone numbers in use in the new geographic area are assigned the new area code.

Geographic splits divide cities, neighborhoods, and local governments, forcing callers to learn the geographic boundaries of the new and old area codes.

The telecommunications industry is working on a way to forestall the creation of new area codes. It's possible to use existing numbers more efficiently by splitting the 10,000-number blocks associated with each telephone prefix into smaller segments.

The 10,000-number block restriction means, for instance, that if a city needs 23,000 telephone numbers, it's required to have three 10,000-block prefixes, leaving 7,000 telephone numbers unused. If the block could be split, a nearby community could use the excess numbers, keeping that prefix in the geographic area. The technology doesn't exist right now to split the 10,000 number blocks.

Area Code Change Checklist

1. Find out whether you have to reprogram your PBX to be able to direct-dial to new area codes. Set up a schedule with your vendor to update your phone system's software, remotely if possible, as area codes change.

2. Fax your overseas customers about new area codes and when they can use them. Suggest they contact an international operator if they have any trouble reaching you.

3. If you have branches overseas, make sure the phone companies serving those branches are aware of area code changes.

4. Inform your customers in advance what your new area code will be. Don't risk losing business because people can't find you.

For More Area Code Information

Tele-tech services has a current list of area codes by number, by state and a schedule for new area code implementation at www.tariffs.com. They also have a great newsletter that comes via US Mail or email.

To get info on specific new area codes, check www.bellcore.com/
NANP/XXX.html, where XXX is the new area code.

Bellcore's Web pages tell you what the new area code replaces,
lists new area code's test number and refers you to a help line to
call if there are any problems. They also tell you when and where
the area code goes into effect.

Area Code List

In Numerical Order

(subject to change)

201	New Jersey, NE	219	Indiana, northern
202	District of Columbia	224	Illinois, far NE, Chicago
203	Connecticut, SW	225	Louisiana, central eastern
204	Manitoba		
205	Alabama, northern	228	Mississippi, SE
206	Washington, Seattle area	231	Michigan
207	Maine	240	Maryland, western and southern
208	Idaho		
209	California, N central	242	Bahamas
210	Texas, San Antonio area	246	Barbados
212	New York, Manhattan	248	Michigan, Oakland County
213	California, Los Angeles		
214	Texas, Dallas	250	British Columbia, all except Vancouver
215	Pennsylvania, Philadelphia		
		252	North Carolina, NE
216	Ohio, Cleveland, NE	253	Washington, Tacoma and S Seattle
217	Illinois, central		
218	Minnesota, northern	254	Texas, N central

256 Alabama, northern
      except Birmingham
264 Anguilla
267 Pennsylvania, SE
268 Antigua and Barbuda
281 Texas, suburban Houston
284 British Virgin Islands
301 Maryland, NW
302 Delaware
303 Colorado, Denver area
304 West Virginia
305 Florida,
      SW including Miami
306 Saskatchewan
307 Wyoming
308 Nebraska, western
309 Illinois, upper NW
310 California,
   | Santa Monica area
312 Illinois, Chicago
313 Michigan, Detroit area
314 Missouri, St Louis area
315 New York, Syracuse
      and N central NY
316 Kansas, southern
317 Indiana, Indianapolis area
318 Louisiana,
      western and northern
319 Iowa, eastern
320 Minnesota, central
      except Minneapolis
323 California, around
      central Los Angeles
330 Ohio, central NE
334 Alabama, southern
336 North Carolina, NW

340 US Virgin Islands
345 Cayman Islands
352 Florida, NW central
360 Washington, western
      except Seattle
401 Rhode Island
402 Nebraska, eastern
403 Alberta, southern
404 Georgia, Atlanta area
405 Oklahoma, central
      including Oklahoma City
406 Montana
407 Florida, NE central
408 California, south bay
      (San Jose) area
409 Texas, SE except
      Houston area
410 Maryland, NE including
      Baltimore
412 Pennsylvania,
      Pittsburgh area
413 Massachusetts, western
414 Wisconsin, E central
415 California, San Francisco
      and N bay area
416 Ontario, Toronto area
417 Missouri SW
418 Quebec, northern
419 Ohio,
      NW including Toledo
423 Tennessee, eastern
424 California,
      W of Los Angeles
425 Washington, N of Seattle
435 Utah, all except Ogden-
      Salt Lake-Provo area

440 Ohio, NE except
Cleveland
441 Bermuda
443 Maryland, eastern
including Baltimore
450 Quebec, S central
except Montreal
456 International Inbound
469 Texas, Dallas area
473 Grenada
484 Pennsylvania, SE
500 Personal Comm. Services
501 Arkansas, NW including
Little Rock
502 Kentucky, western
503 Oregon, NW including
Portland & Salem
504 Louisiana, SE including
New Orleans
505 New Mexico
506 New Brunswick
507 Minnesota, southern
508 Massachusetts,
southern and SE
509 Washington, eastern
510 California, Oakland &
inland N bay area
512 Texas,
Austin and S central
513 Ohio, SW
including Cincinnati
514 Quebec, Montreal area
515 Iowa, central (N to S)
516 New York, Long Island
517 Michigan,
central (N to S)

518 New York, NE
519 Ontario, SW
520 Arizona, except Phoenix
530 California,
NE except Sacramento
540 Virginia,
western and northern
541 Oregon, all but far NW
559 California,
inland S central
561 Florida, eastern central
(West Palm)
562 California, southern
(Long Beach area)
570 Pennsylvania, NE
573 Missouri,
eastern except St Louis
580 Oklahoma, SW
600 Canada/Services
601 Mississippi,
Jackson area
602 Arizona, Phoenix area
603 New Hampshire
604 British Columbia,
Vancouver area
605 South Dakota
606 Kentucky, eastern
607 New York, S central
608 Wisconsin,
SW including Madison
609 New Jersey, southern
610 Pennsylvania,
Philadelphia suburbs
611 Repair service
612 Minnesota,
Minneapolis area

613 Ontario,
SE including Ottawa
614 Ohio, Columbus area
615 Tennessee, central
including Nashville
616 Michigan,
western third of state
617 Massachusetts,
Boston area
618 Illinois, central
619 California, central
San Diego
626 California, Pasadena area
630 Illinois, E central and
Chicago suburbs
649 Turks & Caicos islands
650 California, S bay area
651 Minnesota, St Paul area
660 Missouri,
N except KC and St Jo
661 California, S central
including Bakersfield
664 Montserrat
670 Commonwealth of
Northern Mariana Islands
671 Guam
678 Georgia, Atlanta area
700 IC Services
701 North Dakota
702 Nevada,
Las Vegas area and S
703 Virginia,
DC & Arlington areas
704 North Carolina,
S central
705 Ontario, NE

706 Georgia, N except
concentric Atlanta zones
707 California,
coastal N of bay area
708 Illinois, S Chicago area
709 Newfoundland
710 U.S.Government
712 Iowa, western
711 TRS access
713 Texas, central Houston
714 California,
southern including
N Orange County
715 Wisconsin, northern
716 New York, western
including Buffalo
717 Pennsylvania, central
including Harrisburg
718 New York, boros other
than Manhattan
719 Colorado, SE
720 Colorado, Denver area
724 Pennsylvania,
SW except Pittsburgh
727 Florida, W central
732 New Jersey,
E central costal
734 Michigan, extreme SE
740 Ohio, large SE area
except Columbus
757 Virginia, extreme SE
758 St. Lucia
760 California, large SE area
765 Indiana, central
except Indianapolis
767 Dominica

770 Georgia,
suburban Atlanta

773 Illinois, Chicago area
(not central)

775 Nevada,
except Las Vegas and S

780 Alberta, N including
Edmonton

781 Massachusetts,
Boston suburbs

784 St. Vincent & Grenadines

785 Kansas, northern

786 Florida,
SE including Miami

787 Puerto Rico

800 Toll Free Services

801 Utah, Salt Lake City,
Provo, Ogden areas

802 Vermont

803 South Carolina, central
including Columbia

804 Virginia,
central and mid S
includes Richmond

805 California,
mid-S coastal area
including Santa Barbara

806 Texas, panhandle
including Amarillo

807 Ontario, western

808 Hawaii

809 Caribbean Islands

810 Michigan, central eastern
except Detroit area

812 Indiana, southern
including Evansville

813 Florida, central west
coast including Tampa

814 Pennsylvania,
NW to mid-S including
State College

815 Illinois, northern
except Chicago area

816 Missouri, Kansas City
& St Joseph area

817 Texas,
N central, W of Dallas,
including Ft Worth

818 California,
N of Los Angeles
including Burbank

819 Quebec, southern half
except Montreal area

828 North Carolina,
SW including Asheville

830 Texas, mid SW border

831 California,
central coast including
Monterey and inland

832 Texas, Houston

843 South Carolina,
coastal third of state

847 Illinois,
NE corner including
Chicago suburbs

850 Florida,
W part of panhandle

858 California, S includes
northern San Diego

860 Connecticut,
all but SW,
includes Hartford

864 South Carolina,
   NW third of state
867 Yukon/NW Territories
868 Trinidad/Tobago
869 St.Kitts/Nevis
870 Arkansas, all except large
   area around Little Rock
876 Jamaica
877 Toll Free *800) Service
880 Paid 800 Service
881 Paid 888 Service
882 Paid 877 Service
888 Toll Free Service
900 900 Service
900 900 Service
901 Tennessee,
   W including Memphis
902 Nova Scotia/
   Prince Edward Island
903 Texas, NE
   except Dallas area
904 Florida, N Atlantic
   including Jacksonville
905 Ontario,
   central SE
   border except Toronto
906 Michigan, upper
907 Alaska
908 New Jersey, mid N
909 California, S interior
   including San Bernadino
910 North Carolina,
   SE including Fayetteville
912 Georgia, southern
913 Kansas, NE including
   Kansas City area

914 New York, SE including
   Albany but not NYC
915 Texas, central W
   including El Paso
916 California,
   Sacramento area
917 New York, boros other
   than Manhattan
918 Oklahoma, NE
919 North Carolina, NE
   central including Raleigh
920 Wisconsin, SE except
   Milwaukee area
925 California,
   inland E bay area
931 Tennessee, central
   except Nashville area
935 California, far SW
   including S San Diego
937 Ohio, central western
940 Texas,
   N central E of panhandle
941 Florida, SW
   including Ft Myers
949 California, coastal
   below Los Angeles
   includes Irvine
954 Florida, coastal SE
   including Ft Lauderdale
956 Texas, SW edge
970 Colorado, all northern
   and western areas
972 Texas, Dallas area
973 New Jersey, NW
978 Massachusetts,
   N except Boston

**In Alphabetical Order
By State**

900 900 Service
900 900 Service
205 Alabama, northern
256 Alabama, northern
     except Birmingham
334 Alabama, southern
907 Alaska
403 Alberta, southern
780 Alberta,
     N including Edmonton
264 Anguilla
268 Antigua and Barbuda
520 Arizona, except Phoenix
602 Arizona, Phoenix area
501 Arkansas, NW
     including Little Rock
870 Arkansas,
     all except large area
     around Little Rock
242 Bahamas
246 Barbados
441 Bermuda
250 British Columbia,
     all except Vancouver
284 British Virgin Islands
604 British Columbia,
     Vancouver area
209 California, N central
213 California, Los Angeles
310 California,
     Santa Monica area
323 California, around
     central Los Angeles

408 California, south bay,
     San Jose area
415 California, San Francisco
     and N bay area
424 California,
     W of Los Angeles
510 California, Oakland &
     inland N bay area
530 California, NE except
     Sacramento
559 California,
     inland S central
562 California, southern,
     Long Beach area
619 California,
     central San Diego
626 California, Pasadena area
650 California, S bay area
661 California, S central
     including Bakersfield
707 California,
     coastal N of bay area
714 California, southern,
     Orange County
760 California, large SE area
805 California,
     mid-S coastal area
     including Santa Barbara
818 California,
     N of Los Angeles
     including Burbank
831 California, central coast
     including Monterey
     and inland
858 California, S includes
     northern San Diego

909 California, S interior
including San Bernadino
916 California,
Sacramento area
925 California,
E bay area
935 California, far SW
including S San Diego
949 California, coastal
below Los Angeles
includes Irvine
600 Canada/Services
809 Caribbean Islands
345 Cayman Islands
303 Colorado, Denver area
719 Colorado, SE
720 Colorado, Denver area
970 Colorado, all northern
and western areas
670 Commonwealth of
Northern Mairana Islands
203 Connecticut, SW
860 Connecticut, all but SW,
includes Hartford
302 Delaware
202 District of Columbia
767 Dominica
305 Florida,
SW including Miami
352 Florida, NW central
407 Florida, NE central
561 Florida, eastern central
including West Palm
727 Florida, W central
786 Florida,
SE including Miami

813 Florida,
central west coast
including Tampa
850 Florida,
W part of panhandle
904 Florida, N Atlantic
including Jacksonville
941 Florida,
SW including Ft Myers
954 Florida, coastal SE
including Ft Lauderdale
404 Georgia, Atlanta area
678 Georgia, Atlanta area
706 Georgia, N except
concentric Atlanta zones
770 Georgia,
suburban Atlanta
912 Georgia, southern
473 Grenada
671 Guam
808 Hawaii
700 IC Services
208 Idaho
217 Illinois, central
224 Illinois,
far NE, Chicago
309 Illinois, upper NW
312 Illinois, Chicago
618 Illinois, central
630 Illinois, E central,
Chicago suburbs
708 Illinois, S Chicago area
773 Illinois, Chicago area
(not central)
815 Illinois, northern except
Chicago area

847 Illinois,
    NE corner including
    Chicago suburbs
219 Indiana, northern
317 Indiana, Indianapolis area
765 Indiana, central except
    Indianapolis
812 Indiana, southern
    including Evansville
456 International Inbound
319 Iowa, eastern
515 Iowa, central (N to S)
712 Iowa, western
876 Jamaica
316 Kansas, southern
785 Kansas, northern
913 Kansas, NE including
    Kansas City area
502 Kentucky, western
606 Kentucky, eastern
225 Louisiana, central eastern
318 Louisiana,
    western and northern
504 Louisiana, SE including
    New Orleans
207 Maine
204 Manitoba
240 Maryland,
    western and southern
301 Maryland, NW
410 Maryland,
    NE including Baltimore
443
    Maryland, eastern
    including Baltimore
413 Massachusetts, western

508 Massachusetts,
    southern and SE
617 Massachusetts,
    Boston area
781 Massachusetts,
    Boston suburbs
978 Massachusetts,
    N except Boston
248 Michigan,
    Oakland County
313 Michigan, Detroit area
517 Michigan,
    central (N to S)
616 Michigan,
    western third of state
734 Michigan, extreme SE
810 Michigan, central eastern
    except Detroit area
906 Michigan, upper
218 Minnesota, northern
320 Minnesota, central
    except Minneapolis
507 Minnesota, southern
612 Minnesota,
    Minneapolis area
651 Minnesota, St Paul area
228 Mississippi, SE
601 Mississippi, Jackson area
417 Missouri SW
314 Missouri, St Louis area
573 Missouri,
    eastern except St Louis
660 Missouri, N except
    KC and St Jo
816 Missouri, Kansas City
    & St Joseph area

406 Montana
664 Montserrat
308 Nebraska, western
402 Nebraska, eastern
702 Nevada,
     Las Vegas area and S
775 Nevada, except
     Las Vegas and S
506 New Brunswick
603 New Hampshire
201 New Jersey, NE
609 New Jersey, southern
732 New Jersey,
     E central costal
908 New Jersey, mid N
973 New Jersey, NW
212 New York, Manhattan
315 New York, Syracuse and
     N central NY
516 New York, Long Island
518 New York, NE
607 New York, S central
716 New York,
     western including Buffalo
718 New York, boros
     other than Manhattan
914 New York, SE including
     Albany but not NYC
917 New York, boros
     other than Manhattan
505 New Mexico
709 Newfoundland
252 North Carolina, NE
336 North Carolina, NW
704 North Carolina,
     S central

828 North Carolina,
     SW including Asheville
910 North Carolina,
     SE including Fayetteville
919 North Carolina, NE
     central including Raleigh
701 North Dakota
902 Nova Scotia/
     Prince Edward Island
216 Ohio, Cleveland, NE
330 Ohio, central NE
419 Ohio,
     NW including Toledo
440 Ohio,
     NE except Cleveland
513 Ohio, SW including
     Cincinnati
614 Ohio, Columbus area
740 Ohio, large SE area
     except Columbus
937 Ohio, central western
405 Oklahoma, central
     including Oklahoma City
580 Oklahoma, SW
918 Oklahoma, NE
416 Ontario, Toronto area
519 Ontario, SW
613 Ontario,
     SE including Ottawa
705 Ontario, NE
807 Ontario, western
905 Ontario, mid SE border
     except Toronto
503 Oregon, NW including
     Portland & Salem
541 Oregon, all but far NW

880 Paid 800 Service
881 Paid 888 Service
882 Paid 877 Service
215 Pennsylvania,
    Philadelphia
267 Pennsylvania,
    Philadelphia
412 Pennsylvania,
    Pittsburgh area
484 Pennsylvania, SE
570 Pennsylvania, NE
610 Pennsylvania,
    Philadelphia suburbs
717 Pennsylvania, central
    including Harrisburg
724 Pennsylvania,
    SW except Pittsburgh
814 Pennsylvania,
    NW to mid-S
    including State College
500 Personal Comm. Services
787 Puerto Rico
418 Quebec, northern
450 Quebec, S central
    except Montreal
514 Quebec, Montreal area
819 Quebec, southern half
    except Montreal area
611 Repair service
401 Rhode Island
758 St. Lucia
784 St. Vincent & Grenadines
869 St.Kitts/Nevis
306 Saskatchewan
803 South Carolina, central
    including Columbia

843 South Carolina,
    coastal third of state
864 South Carolina,
    NW third of state
605 South Dakota
423 Tennessee, eastern
615 Tennessee, central
    including Nashville
901 Tennessee,
    W including Memphis
931 Tennessee, central
    except Nashville area
210 Texas, San Antonio area
214 Texas, Dallas
254 Texas, N central
281 Texas, suburban Houston
409 Texas,
    SE except Houston area
469 Texas, Dallas area
512 Texas,
    Austin and S central
713 Texas, central Houston
806 Texas, panhandle
    including Amarillo
817 Texas,
    N central, W of Dallas,
    including Ft Worth
830 Texas, mid SW border
832 Texas, Houston
903 Texas, NE except
    Dallas area
915 Texas, central W
    including El Paso
940 Texas,
    N central E of panhandle
956 Texas, SW edge

972 Texas, Dallas area
800 Toll Free Services
877 Toll Free (800) Service
888 Toll Free Service
868 Trinidad/Tobago
711 TRS access
649 Turks & Caicos islands
710 U.S.Government
340 U.S. Virgin Islands
435 Utah, all except Ogden-Salt Lake-Provo area
801 Utah, Salt Lake City, Provo, Ogden areas
802 Vermont
540 Virginia, western and northern
703 Virginia, DC & Arlington areas
757 Virginia, extreme SE

804 Virginia, central and mid S includes Richmond
206 Washington, Seattle area
253 Washington, Tacoma and S Seattle
360 Washington, western except Seattle
425 Washington, N of Seattle
509 Washington, eastern
304 West Virginia
414 Wisconsin, E central
608 Wisconsin, SW including Madison
715 Wisconsin, northern
920 Wisconsin, SE except Milwaukee area
307 Wyoming
867 Yukon/NW Territories

Tele-Tech Services keeps an updated list of area codes and a schedule of new area code implementation dates at http://www.tariffs.com.

To get an update on area codes in general, check out Bellcore's Web page on the subject at www.bellcore.com/NANP/

To get info on specific new area codes, check www.bellcore.com/NANP/XXX.html. Put the area code you're interested in where XXX is.

Analog Transmission, Multiplexing and Circuits

Frequency Division Multiplexing, Central Office Trunks, Analog Private Lines, Tie Lines and Direct Inward Dial

Analog comes from the word "analogous," which means "similar to." In telephone transmission, the signal being transmitted-voice, video, or image-is "analogous" to the original signal. The only difference between a signal being transmitted and the real thing is that the electrically transmitted signal is at a higher frequency than regular voice.

In correct English usage, "analog" is meaningless as a word by itself. But in telecommunications, analog means telephone transmission and/or switching that is not digital.

Standard telephone signals have historically been analog transmissions that vary continuously, representing the relative loudness in the speech they transmit. Analog signals can take on any value between rather wide limits.

Frequency Division Multiplexing (FDM)

Within a few decades after the telephone was commercialized, the number of local loops and interoffice trunks increased rapidly. The skies above cities were filled with poles, cross arms, and countless wires. Eventually there was no more room for additional wire.

The invention of the vacuum tube, plus amplifiers and oscillators, enabled telco engineers to develop frequency division multiplexing.

Multiplexing reduced the need for wires. The FDM (Frequency Division Multiplexing) technique assigns each voice channel to a different 4 kHz (4,000 Hertz) frequency band. When added together, 24 voice channels total 96 kHz (96,000 Hertz), well within the capacity of a copper twisted pair.

A solid copper loop can carry a much wider bandwidth signal than the 4 kHz required for a single voice conversation; a pure copper path can carry frequencies above 1 MHz (1,000,000 Hertz).

In the early stage of vacuum tube designs, however, conservative practice led to a top figure of 96 MHz. This allowed 24 voice channels of 4 kHz each to be multiplexed on two pairs of wire (one for send and one for receive).

Channel 1 is transmitted at between zero and 4 kHz. Channel 2 is translated electronically to the frequency range 4 to 8 kHz. Channel 3, becomes 8 to 12 kHz and so on. Adding all channels together produced a broadband signal that can be treated as one channel for transmission over interoffice trunks.

FDM was still in use in the 1980s for long-haul circuits over microwave transmission links, but has since largely been replaced with digital facilities. FDM gave way to digital signal processing because of its susceptibility to noise. Digital transmission also alleviates the high cost of switching, controlling, and maintaining physical channels.

Central Office Trunks

The two-wire electrical path between the phone company central office and the customer is called a central office trunk. (A four-wire or two-pair path is called a circuit.)

> ### See the Data Communications, Digital Voice and the OSI Model chapter for more about digital transmission of analog signals.

Central Office (CO) trunks have individual phone numbers that are used to identify them for incoming and outgoing calls.

CO trunks are the cheapest way to get dial tone in most places, and still the way to go for most SOHO and small businesses that don't need high-speed Internet or data access.

Each trunk is connected to the central office by two wires called a wire pair or the local loop. One of the wires is called T, for tip, and the other is called R for ring, referring to the tip and ring parts of the plug used in the early manual switchboards.

CO trunks are available from the telephone company in many flavors, inbound-only, outbound-only or bothway, and with or without features like call waiting, call forward busy, call forward RNA (ring no answer).

The best way to set up analog trunks for a key system (if you can justify a PBX, you'll probably use digital trunks) is to have your incoming trunks designated as a hunt group. Your primary trunk (usually the phone number that's listed) forwards to a second trunk when busy, the second hunts to the third when busy, and so on.

It is also good idea to order your trunks with RNA forwarding from one to the next. If one of the trunks in your hunt group goes dead, the callers that hit that trunk will just hear incessant ringing unless the trunk forwards to the next if it's not answered. If the trunks don't forward RNA, the caller won't get through.

Loop vs Ground Start Trunks

Loop start trunks are powered by the phone company. In the United States, the voltage applied to the line to drive the telephone is 48

Volts Direct Current (VDC); some countries use 50 Volts. When a CO trunk that is loop start is taken off the hook, the line voltage drops from 48 Volts to between 9 and 3 Volts, depending on the length of the loop. The loop of current is complete when the device at the other end goes off-hook. The change in the flow of current acts as a signal to the exchange equipment to "handle" a new call. Dial tone indicates that the exchange is ready to accept your call.

Central Office voltage is direct current (DC) supplied by lead acid cells, for a hum-free supply and as assurance against electric company failures.

Ground start trunks are not provided with power for the communications signal or for ringing. PBXs, and recently, key systems, use ground start trunks preferentially to provide signaling from on-premises. This gives your phone system more control over the trunk, and is why the dial tone and ringing cadences on PBX phones sound different than those from standard analog single line sets.

Analog Private Lines

Analog private line services have been sold for years (even decades) and are an imbedded, legacy market.

Although analog private line services are technologically obsolete, voice grade analog services will be around for many years to come. There is always technological inertia in communications markets, and it takes a long time for the demand to fall to zero for any product.

Tie Lines

Tie lines are dedicated "dry" (no dial tone or voltage) circuits between two points; a tie trunk is a dedicated circuit linking two switches. You can use tie lines to connect two distant offices, for voice and data. Voice tie lines directly connect two PBXs that can then be programmed for one or two digital access code.

Trunks between analog PBXs most often terminate individually using an "E&M" interface. E and M are signaling leads that accompany a voice path of 2 or 4 wires. A PBX seizes a tie trunk in one of several ways, typically by grounding the M lead (which appears as the E lead at the far end).

Each PBX sends signals on its M lead and looks for signals on its E lead (remembered easily as Ear and Mouth, from the switch's point of view).

Tie lines were used frequently before stiff competition in long distance rates drove prices down because the monthly flat rate of tie lines was cheaper at a certain volume of calls than per-minute long distance service. Now tie lines are back again, running digital data/voice services.

> **See the Frame Relay, ATM and xDSL chapters for more about digital voice using tie lines.**

Direct Inward Dial (DID)

DID (Direct Inward Dialing) associates multiple direct-dial phone numbers with a fewer number of trunks. DID is generally used so that individual extension numbers on PBX phones can be direct-dialed from outside, but DID is also used for fax routing.

The Central Office switch knows which DID trunks are associated with which DID phone numbers. When the CO receives a call for one of your DID numbers, it switches it to one of your DID trunks.

Your telephone system understands the signaling information that accompanies the DID call and can switch it internally to the phone that matches the DID extension that was dialed.

The phone company rents DID phone numbers to you in blocks

of 20 or 25 and also charges you (about twenty percent more than standard business line rates) monthly for the trunks. There are some setup and installation costs.

Using DID for Fax

DID is also used with fax machines, modems, boards and servers to route inbound fax calls.

Calls to separate DID numbers are all sent in via the same DID trunk. The fax board (or external DID decoder box) decodes the signal from the telco central office that indicates which number was dialed and uses this number to route the fax to the appropriate user or department.

This is a less expensive way to have a lot of fax machines that can be dialed from outside, and, most importantly, the messages for extensions (or machines) can be forwarded from one to another. When one fax machine is busy, the call moves to the next one.

Summary

Analog voice services are alive and kicking for systems up to about 50 telephones. They're usually cheaper for smaller systems than digital services, and are easy to order and install. DID services, in particular, make smaller firms look like big enterprises by allowing every employee a direct-dial number.

Signaling and ANI

In-band vs. out-of-band, DTMF and ANI

The telephone network is a widely distributed system of intelligent switches. These switches, to establish and tear down calls, must communicate with each other and with the end users' equipment. The communication process is called signaling.

The original signaling was the human voice, when a caller spoke to the operator. The first automatic terminal equipment used special switch-generated call progress tones to give call status information to callers.

Rotary signaling turns current on and off corresponding to the number dialed. As the dial is released after each digit of a phone number, the current is turned off and on from one to 10 times, depending on the digit. This "coded" signal goes to the Central (telephone switching) Office and used to make the correct connection to the destination telephone number.

An analog input device (like a phone) takes the acoustic signal (which is a natural analog signal) and converts it into an electrical equivalent in terms of volume (signal amplitude) and pitch (frequency of wave change).

Removing the handset at the ringing telephone results in a loop current flow. When a called party removes the handset in response to a ring, the loop to that phone is completed by its closed switchook and loop current flows through the called telephone. The central office then removes the ringing signal and ringback tone from the circuit.

> **See the PBX (Private Branch Exchange) chapter for more about ring generators and progress tones.**

Dial Tone Multi Frequency (DTMF)

Dual Tone Multi-Frequency (DTMF), makes possible touch-tone or "pushbutton" dialing, and is the best-known in-band terminal equipment-to-network signaling. The equipment at the customer end generates pairs of tones to represent each dialed digit.

Bell Labs developed DTMF in order to have a dialing system that could travel across microwave links and work rapidly with computer controlled exchanges.

A touch tone keypads' 12 keys produce a varying electrical signals when pressed. The sounds are each a combination of two of seven possible tones and are converted into electrical signals that are sent to the CO (Central Office). At the exchange, the signals are recognized as numbers and the call is switched.

Each transmitted digit consists of two separate audio tones that sound together. The three vertical columns on the keypad (there's a 4th column to indicate the urgency of the call on government Autovon phones) are known as the high group because the tones are all above 1200 Hz. The four horizontal rows are "the low group" because the tones are all below 1000 Hz. For example, the digit 8 is a combination of 1336 Hz and 852 Hz. The two tones of each number are within 3 dB (decibels) of each other in amplitude. (The telephone company calls a distortion of this amplitude ratio "Twist".)

The * sign is usually called "star" or "asterisk." The # sign, often referred to as the "pound sign" is actually called an octothorpe. These digits are used for controlling answering machines, bringing up remote databases and repeater control.

A DTMF digit must be sounded for at least 100 milliseconds in order to be decoded.

In-Band vs Out-of-Band Signaling

Signaling can take place in-band, through same channel as the voice or data conversation, or out-of-band, through some communication channel other than the talk path, like a separate wire or a dedicated data channel.

Traditional analog telephone signaling is in-band. You hear dial tone, dialed digits, and ringing/busy/reorder tones over the same voice channel on the same pair of wires. When the call is completed, you talk over the same path that was used for the call setup and signaling.

Analog interoffice trunks formerly used strictly in-band signaling. The signals to set up a call between one switch and another always took place over the same trunk that would eventually carry the call. Switch-to-switch signaling used a series of multifrequency (MF) tones; touch tone dialing between switches.

The D channel used in ISDN services is out-of-band. The proprietary digital signaling used between PBXs and their digital telephone sets is also generally out-of-band.

The switch-to-switch signaling protocol called Signaling System 7 (SS7) used in public and large private telephone networks is out-of-band, as are the links between PBXs and computer telephony (CT) systems.

Out-of-band signaling establishes a separate digital channel for the transmission of signaling information, so, for a given bandwidth, it allows data transport at higher speeds than does in-band signaling. Out-of-band signaling can also take place at any time during the call and enables signaling information to be flexibly switched, across a network if necessary, without a direct circuit connection. This frees the information about the call from the call itself.

ANI (Automatic Number Identification)

> **What is DNIS?**
> Dialed Number Identification Service (DNIS) gives the number dialed by the caller to the equipment that answers. An example of DNIS use is 800 ordering numbers for several catalogs being serviced by the same pool of agents who, with DNIS, see which number was dialed and can answer their phones with greetings from the appropriate catalog company.

Automatic Number Identification (ANI) is a phone company signal that gives you the telephone number of the calling party. The calling party station ID (phone number) can be delivered digitally either in-band, using dual-tone multifrequency (DTMF) or multifrequency (MF) signals, or out-of-band using ISDN PRI, as a part of the call set-up data.

ANI information can be used to route calls, to log specific callers into billing databases, indicate privileges and restrictions, automate order entry transactions, match callers to geographic coordinates, and bring up appropriate customer records from a database when an agent answers the phone. If the caller is known to the company, ANI is capable of providing a name to go with the number so that the agent can greet the caller by name, a nice touch.

(Caller ID offers no frills beyond the number of the caller, and is usually provided by the local telco.)

ANI service offerings from long distance companies include MCI's 800 Enhanced Service Package (ESP), Sprint's Real Time ANI and AT&T's INFO 2. Actual ANI transmission formats differ slightly from carrier to carrier, so make sure the system you're using to capture the ANI information is capable of reading your provider's ANI data stream.

There are two basic designs for ANI systems, stand-alone and switch adjunct. The system you use depends on whether you already have a PBX, and whether the calls need to be switched somewhere else after the ANI information is collected.

A stand-alone ANI system has to be the last communication with your caller because stand-alone systems are incapable of switching calls anywhere else (to an operator or a voicemail system). The most familiar stand-alone ANI systems include FOD (Fax on Demand) document delivery systems and wide-area opinion polling response recorders.

Switch adjunct ANI systems operate in tandem with an associated phone system. The ANI system is installed either in front of or behind the switch. Switch adjunct systems can be used in call centers so the incoming caller is routed to the best (or most familiar) agent.

Switch adjunct ANI systems interface to digital PBXs, collect ANI information and send it to a database. A detailed record, including the caller's phone number, the inbound trunk number over which the call was carried and the call's duration is kept for each inbound call. ANI loggers save raw ANI data in order to analyze calling patterns. An ANI logger can be used for security and fraud control, billing verification backup, real-time channel activity analysis and database creation.

Installed PBXs that aren't ANI-equipped will require an ANI converter. ANI converters are made by PBX companies so that digital switches can interface to ANI-based lines.

The ANI converter extracts the ANI signal sent from the CO (Central Office) and translates it in a format the PBX recognizes. An ANI converter can, at once, capture the ANI information for use by an adjunct identification system, and reformat the last four digits of a DNIS signal into extension numbers for PBX routing.

Data Communications, Digital Voice and the OSI Model

Signaling, Digital Voice, Analog to Digital Conversion, Pulse Code Modulation (PCM), SNA and SDLC and Signaling System 7 (SS7)

The term "bit" is a contraction of binary digit, the smallest unit of digital information-either a zero or a one-a signal for either on or off. The major computer codes use either seven plus a parity bit for error control (ASCII) or eight (EBCDIC) bits to represent one letter, number or symbol.

Each data digit is called a bit. A string of bits, usually eight or 16, that a computer can address as an individual group is called a byte. (A four-bit byte is a nibble.)

The slowest digital transmission speeds are measured in bits per second (bps). Faster communications run at kbps (kilobits per second) or thousands of bits per second, Mbps (megabits per second) or millions of bits per second and Gbps (gigabits per second) or billions of bits per second.

Digital Voice: Analog to Digital Conversion

Every wire acts as an antenna. Copper wires pick up signals from electrical radiation sources that exist almost everywhere. Fluorescent lights, motor starters, photocopiers and ring voltage bleeding over from close copper pairs can cause interference with a signal.

Analog signals fade with distance and must be amplified periodically. (Analog signal amplifiers are spaced at intervals of 18,000-meters.) The noise that is always present on copper circuits is amplified right along with the signal. Eventually the volume of the noise exceeds the signal volume and conversation becomes difficult or impossible to understand. To bring the problem of noise under control, the telephone industry developed digital transmission techniques.

The problem of digital voice comes down to selecting the "best" way to represent a smoothly varying quantity. A limited number of bits can be transmitted most economically. "Best" often implies maintaining voice quality, but there are trade-offs between quality, bandwidth, cost, availability and reliability.

The great difference between analog and digital is that the digital receiver knows to a great extent what to expect. An analog signal legitimately can be any value in a range, so a slight variation due to noise is difficult to detect and almost impossible to remove.

A digital signal, however, can have only a few true values. Deviations from the allowed states represent fading or noise, so a digital amplifier can distinguish between the signal and noise. The line amplifier examines the input at periodicaly, and decides what value the input probably represents. Exactly that value is then sent on the output side.

Narrowband vs Broadband

Narrowband is a comparatively thin slice of bandwidth for transmission of voice and data signals. At operational speeds of 2.6 kbps to 1.54 Mbps, it is most often known as T-Carrier, carrying up to 24 DS0 channels at 64 kbps each.

North America's definition of narrowband, derived from the Bell System "T-carrier" trunking hierarchy, is not the same as the rest of the world's. In Europe, Asia, Latin America and Africa, narrowband has an upper limit of 2.0 Mbps, not 1.54 Mbps, and is referred to as E1, not T1.

Wireless and satellite data service-providers have sometimes labeled their products "broadband" when they are not, and some packet-switching technologies originally designed for broadband, like frame relay, often run at access speeds under T-1 and should be considered narrowband.

Services that push speeds between T-1 and T-3 (45 Mbps) are generally called wideband or broadband services.

Network Architecture

Systems Network Architecture (SNA)
The Systems Network Architecture (SNA) was introduced by IBM in 1974 to provide a framework for joining together their great number of mutually incompatible distributed processing products. SNA was one of the first communications architectures to use a layered model, which later became the basis for the OSI model.

The SNA is a hierarchical network that consists of a collection of machines called nodes. There are four types of nodes; Type 1 (terminals), Type 2 (controllers and machines that manage terminals), Type 4 (front-end processors and machines that take some load off the main CPU) and Type 5 (the main host).

Each node has at least one Network Addressable Unit (NAU). The NAU enables a process to use the network by giving it an address. A process can then reach and be reached by other NAUs.

An NAU can be one of three types; an LU (Logical Unit), a PU (Physical Unit) or an SSCP (System Services Control Point).

Usually there is one SSCP for each Type 5 node and none in the other nodes.

SNA distinguishes five different kinds of sessions: SSCP-SSCP, SSCP-PU, SSCP-LU, LU-LU and PU-PU.

The SSCP (PU Type 5) is usually used in IBM mainframe machines that use channels to connect to control devices such as disks, tapes and communication controllers. These are high speed communications links (up to 17 Mbps).

The communication controller (the FEP Front End Processor, PU type 4) is used to connect low speed SDLC lines. All together the SSCPs, FEPs, channels and SDLC lines connecting them create the SNA backbone. Using SDLC, the FEPs also connect Token Ring LAN or X.25 links and other types of SNA devices such as cluster controllers and RJE stations. RJE stations are PU type 2/2.1 devices and are used to manage LUs (elements such as the display terminal, the 3270 family) which are the endpoint of SNA network.

SNA information may be transmitted within various protocols. Two protocols which are often used to carry SNA information are SDLC and QLLC (which carries SNA information over X.25).

Sychronous Data Link Control (SDLC)
The SDLC (Synchronous Data Link Control) protocol was developed by IBM to be used as the second layer of the SNA hierarchical network. SNA data is carried within the information field of SDLC frames. SDLC is used on multipoint lines and it can support up to 256 terminal control units or secondary stations per line.

Address field. The address field is used to route frames by identifying the sending and destination stations.

Control field. The field following the Address Field is called the Control Field and serves to identify the type of the frame. In addition, it includes sequence numbers, control features and error tracking according to the frame type.

Poll/Final bit. Every frame holds a one bit field called the Poll/Final bit. In SDLC this bit signals which side is 'talking', and provides control over who will speak next and when. When a primary station has finished transmitting a series of frames, it sets the Poll bit, thus giving control to the secondary station. At this time the secondary station may reply to the primary station. When the secondary station finishes transmitting its frames, it sets the Final bit and control returns to the primary station.

Modes of operation. In SDLC there is the notion of primary and secondary stations, defined simply as the initiator of a session and its respondent. The primary station sends commands and the secondary station sends responses.

FCS. The Frame Check Sequence (FCS) enables a high level of physical error control by allowing the integrity of the transmitted frame data to be checked. The sequence is first calculated by the transmitter using an algorithm based on the values of all the bits in the frame. The receiver then performs the same calculation on the received frame and compares its value to the CRC.

SDLC operates in Normal Response Mode (NRM). NRM is master/slave meaning that only one station may transmit frames at any one time (when permitted to do so). The primary station initiates the session and sends commands. The secondary station sends responses. Full polling is used for all frame transmissions.

Signaling System Seven (SS7)

SS7 is a common channel signaling system developed by the CCITT as an international standard for signaling across voice networks.

THE OSI REFERENCE MODEL

Modern computer networks are designed in a highly structured way. To reduce their design complexity, most networks are organized as a series of layers, each built on top of the other.

The OSI Reference Model is based on a proposal developed by the International Standards Organization (ISO). The model is called ISO OSI (Open systems Interconnection) Reference Model because it deals with connecting open systems - systems that are open for communication with other systems.

The OSI model has seven layers.

7. The Application Layer
The application layer provides services to the applications that are running on the client (end user) machine.

The application layer contains a variety of protocols to deal with the hundreds of incompatible terminal types in the world. This layer defines abstract network virtual terminals, and a piece of application software maps the functions of the network virtual terminal onto the real terminal.

Another application layer function is file transfer. Different systems have different file naming conventions, different ways of representing text lines, and so on. Transferring a file between two different systems requires handling these and other incompatibilities. This work, too, belongs to the application layer, as do electronic mail, remote job entry, directory lookup, and various other general-purpose and special-purpose facilities.

6. The Presentation Layer

Presentation converts the information to a format that the client machine can understand.

The presentation layer performs functions that are requested often enough to support a general solution rather than letting each user solve the communications problems themselves. In particular, unlike all the lower layers, interested in moving bits reliably from here to there, the presentation layer is concerned with the syntax and semantics of the information transmitted.

A typical presentation service is the encoding of data in a standard, agreed-upon way. The job of managing these abstract data structures and converting from the representation used inside the computer to the network standard representation is handled by the presentation layer.

Most user programs exchange information such as people's names, dates, amounts of money, and invoices. This data is represented as character strings, integers, floating point numbers, and data structures composed of several simpler items.

Different computers have different codes for representing character strings, integers and so on. In order to make it possible for computers with different representation to communicate, the data structures to be exchanged can be defined in an abstract way, along with a standard encoding method.

The presentation layer also specifies other aspects of information representation, such as data compression and cryptography.

5. The Session Layer

The session layer allows users to establish sessions between their respective machines. A session allows ordinary data transport, as does the transport layer, but it also provides some enhanced services. A session might be used to allow a user to log into a remote time-sharing system or to transfer a file between two machines.

One of the services of the session layer is to manage dialogue control. Sessions can allow traffic to go in one or both directions at the same time. If traffic is unidirectional, the session layer keeps track of whose turn it is.

A related session service is token management. For some protocols, it is essential that both sides do not attempt the same operation at the same time. To manage these activities, the session layer provides tokens that can be exchanged. Only the side holding the token may perform the critical operation.

Another session service is synchronization. The session layer provides a way to insert checkpoints into the data stream so that after a crash, only the data after the last checkpoint has to be sent again. The session layer andles problems that are not communication issues.

4. The Transport Layer

The transport layer provides end-to-end communication control. The basic function of the transport layer is to accept data from the session layer, split it up into smaller units if need be, pass these to the network layer, ensuring that the pieces all arrive correctly at the other end. Furthermore, all this must be done efficiently, and in a way that isolates the session layer from the inevitable changes in hardware technology.

Under normal conditions, the transport layer creates a distinct network connection for each transport connection required by the session layer. If the transport connection requires a high throughput, the transport layer might create multiple network connections, dividing the data between connections to improve throughput.

If creating or maintaining a network connection is expensive, the transport layer might multiplex several transport connections onto the same network connection to reduce the cost. In all cases, the transport layer is required to make the multiplexing transparent to the session layer.

The transport layer also determines what type of service to provide to the session layer, and ultimately, the users of the network. The most popular type of transport connection is an error-free point-to-point channel that delivers messages in the order in which they were sent. However, other possible kinds of transport, include service, transport of isolated messages with no guarantee about the order of delivery, and broadcasting of messages to multiple destinations. The type of service is determined when the connection is established.

The transport layer is a true source-to-destination or end-to-end layer. A program on the source machine carries on a conversation with a similar program on the destination machine, using the message headers and control messages.

Many hosts are multi-programmed, which implies that multiple connections will be entering and leaving each host. There needs to be some way to tell which message belongs to which connection. The transport header is one place for this information.

In addition to multiplexing several message streams onto one channel, the transport layer musk take care of establishing and deleting connections across the network. This requires some kind of naming mechanism so that a process on one machine has a way of describing with whom it wishes to converse. There must also be a mechanism to regulate the flow of information so that a fast host cannot overrun a slow one. Flow control between hosts is distinct from flow control between switches, although similar principles apply to both.

3. The Network Layer
The network layer routes information in the network, controlling the operation of the subnet. The network layer determines how packets are routed from source to destination. Routes can be based on tables that are "wired into" the network and rarely changed. They can also be determined at the start of each conversation, for not-so-lengthy sessions, like dialup Internet sessions. Or they can be highly dynamic, determined anew for each packet to reflect the current network load.
If too many packets are present in the subnet at the same time, they will get in each other's way, forming bottlenecks. The control of such congestion also belongs to the network layer.

Since the operators of the subnet may well expect remuneration for their efforts, there is often some accounting function built into the network layer. At the very least, the software must count how many packets or characters or bits are sent by each customer to produce billing information. When a packet crosses a national border, with different rates on each side, the accounting can become complicated.

When a packet has to travel from one network to another to get to its destination, many problems can arise. The addressing used by the second network may be different from the first one. The second one may not accept the packet at all because it is too large. The protocols may differ, and so on. It is up to the network layer to overcome all these problems to allow heterogeneous networks to be interconnected.

In broadcast networks, the routing problem is simple, so the network layer is often thin or even nonexistent.

2. The Data Link Layer
The data link layer should provide error control between adjacent network nodes.

The main job of the data link layer is to take a raw transmission facility and transform it into a line that appears free of transmission errors in the network layer.

The sending device breaks the input data it receives into data frames (typically a few hundred bytes), transmits the frames sequentially, and processes the acknowledgment frames when they're sent back by the receiving unit.

Since the physical layer merely accepts and transmits a stream of bits regardless of their structure, it is up to the data link layer to create and recognize frame boundaries. This can be accomplished by attaching special addressing information to the beginning and end of the frames.

1. The Physical Layer
The OSI physical layer connects the hardware to the transmission media. The physical later is concerned with transmitting raw data bits over a communication channel.

The physical layer deals largely with mechanical, electrical, and procedural interfaces, and the physical transmission medium, which lies below the physical layer. The physical transmission medium is the actual wire or fiber that the signal is transmitted across.

Physical layer designs make sure that when one side sends a "1" bit, it is received by the other side as the same single bit.

Physical layer-related decisions include how many volts should be used to represent a "1" and how many for a "0", how long in microseconds a bit lasts, and whether transmission may proceed simultaneously in both directions. Also included in the physical layer is how the initial connection is established,how it is torn down, and the type and pin-out of the network connector.

An out-of band, single channel is used for signaling. The hardware and software functions of the SS7 system are divided into protocol layers that loosely correspond to the OSI 7-layer model.

SS7 protocol layers include the MTP-2 (Message Transfer Part Level 2), the MTP-3 (Message Transfer Part Level 3), the SCCP (Signaling Connection Control Part), the ISUP (ISDN User Part), the TCAP (Transaction Capabilities Application Part), the MAP (Mobile Application Part), and the DUP (Data User Part).

Message Transfer Part - Level 3 (MTP-3) transfers messages between the nodes of the signaling network. MTP-3 ensures reliable transfer of the signaling messages, even when the signaling links and/or transfer points fail.

The protocol includes the appropriate functions and procedures necessary both to inform the remote parts of the signaling network of the consequences of a fault, and appropriately reconfigure the routing of messages through the signaling network.

Message Transfer Part - Level 2 (MTP-2) is a signaling link that works with MTP-3 to reliably transfer signaling messages between two directly connected switches. The Signaling Connection Control Part (SCCP) provides connectionless and connection-oriented network services and also addresses protocol translation. The SCCP enhancements to MTP provide services that correspond to the OSI Network layer 3.

ISUP is the ISDN User Part of SS7. ISUP defines the protocol and procedures used to setup, manage and release circuits that carry voice and data calls over the PSTN.

Mobile Application Part (MAP) messages are sent between mobile (cellular) switches and databases to support equipment and user identification for roaming. When a mobile subscriber roams into a new switching center (MSC) coverage area, the visitor location register requests service profile information from the subscriber's home location register (HLR) using MAP (mobile application part) information . The MAP information is carried within TCAP messages.

Data User Part (DUP) performs call control, facility registration and call cancellation for international common channel signaling for circuit-switched data transmission services.

Internet Standards

SMTP, POP3, IMAP4

Support for Internet messaging standards is vital. There are three standards which are particularly relevant to messaging: SMTP, POP3, and IMAP4.

SMTP is one of the basic building blocks behind Internet email. It is a protocol used to send email to a server on the Internet.

POP3 (Post Office Protocol version 3) is a simple mail retrieval protocol that a desktop program uses to request a list of email messages and then retrieve them. POP3 is also very widely used by Internet email programs, but has one major limitation, it downloads messages from the mail server, or post office, in a single batch. This is fine if you are dealing with small messages, but it can be inconvenient for remote users when several large files are being received..

IMAP4 (Internet Message Access Protocol version 4) is the successor to POP3. It permits the selective retrieval of messages from a post office, and also allows the post office to archive messages for users. This is important because a user may not want to retrieve all of the messages in one batch, and a user may use several different computers to access messages. It's convenient to be able to keep everything in one place.

Integrated Services Digital Network (ISDN)

ISDN is a digital voice and/or data service that (generally) works over existing copper local loops.

ISDN transmits data digitally and, as a result, is less susceptible to static and noise than analog transmissions, although it is very sensitive to outside interference. To avoid interference you have to be within a given distance of the CO (Central Office) equipment that serves you (typically 18,000 feet), and there can't be any other anomalies near the wiring that might interfere with the transmission.

Some applications that ISDN makes possible are telecommuting (remote image sharing and retrieval, inexpensive teleconferencing, combined voice and image collaborations, fast access to remote files or databases), high-speed Internet access, network backup, and LAN-to-LAN or PBX-to-PBX links.

ISDN supports PBX networking to remote workers, plus data networking. You can use ISDN routers to send PBX signaling and functions to remote (telecommuting) employees. If you use web-based resources you can allocate one channel for web data and the other for the PBX functions.

Using an ISDN network bridge or router, It is possible to connect an ISDN line to a LAN so all the PCs on the LAN can share the same ISDN line.

ISDN can substitute for analog leased lines between PBXs or LANs. A break-even between ISDN and leased lines is achieved at about two to three hours per day of usage.

Specialized hardware has evolved to fill each ISDN application niche. These are targeted products, with features well-suited to their designed uses. The ISDN vendors pretty much know what they're selling, and which vendors they're selling against. Pricing is competitive.

NT1s (Network Terminal) and NT2s, also called ISDN adapters, support analog telephone devices like phones, data modems, Group 3 fax machines and answering machines. The NT converts the analog signal into ISDN signals and vice versa.

There are two levels of ISDN service: the Basic Rate (BRI) and the Primary Rate (PRI). Both services include a number of B (bearer) channels and a D (delta) channel. The B channels carry digital data, video, and voice.

The D channel is out-of-band signaling link, carrying packet-switched data using the Signaling System 7 (SS7) protocol, a digital Central Office switch standard. The D channel data can carry caller ID information, advanced maintenance and diagnostic messages, and controls the way the other B channels are dynamically grouped together, so that you can change bandwidth on demand for things like video teleconferencing.

ISDN Basic Rate Interface (BRI)

BRI service provides two B channels (Bearer) and one D channel (Delta), shortened to 2B+D. The B channels each operate at 64 kbps and the D channel operates at 16 kbps. The two B channels can be combined (when they're used together they're said to be BONDed) for speeds of up to 128 Kbs. BRI is used by residential, SOHO and small business customers.

As many as eight separate devices can be connected to the same ISDN BRI circuit, and each one can have more than one telephone number assigned to it. Devices can include network routers and bridges, Group 4 ISDN fax machines, ISDN telephones as well as traditional analog telephones, fax or answering machines.

ISDN is intelligent enough to arbitrate the use of the two B channels between the devices (up to two devices can be in use simultaneously) and route incoming calls to the appropriate device.

ISDN PRI (Primary Rate Interface)

The Primary Rate Interface or PRI in the United States consists of twenty-three 64 kbps B channels and one 64 kbps D channel, called a 23B+D connection. PRI has a total bandwidth of 1.544 Mbps.

In Europe and the Pacific Rim, PRI is a 2.048 Mbps E-1 channel, and consists of either thirty or thirty-one 64 kbps B channels and one 64 kbps D channel, designated 30B+D or 31B+D.

ISDN implementation is still slightly different from nation to nation, but interconnection between any two systems is possible.

ISDN Customer Premises Equipment (CPE)

Terminal Equipment (TE) is any user-owned hardware like a telephone or fax. There are two kinds of terminal equipment, TE1, which refers to equipment that is ISDN compatible, and TE2, which is equipment that is not ISDN compatible

ISDN Devices connect CPE (Customer Premises Equipment) and a network. In addition to fax, telex, PCs, and telephones, ISDN devices may include Terminal Adapters (TAs).

ISDN TERMINAL ADAPTERS (TAs)

TAs connect to your ISDN circuit so that you can have 128 kbps Internet access or data transmission. TAs are not modems, because they don't modulate and demodulate a carrier-wave analog signal, they send digital ones and zeros. Straight-through digital processing means that TAs don't have as much overhead data (which means faster throughput) and don't need very much error checking, so transmission is much faster.

Good ISDN TAs have a built-in analog modem for analog connections and faxing, should support TCP/IP and H.320 video conferencing software/hardware, and should be easy to set up. (Some TAs automatically go out and get all of the switch and connection settings for you from your CO switch.)

A TA, in order to share your ISDN line between ISDN devices or run analog phone equipment on your line, needs a S/T interface for each device. If you're running only one ISDN device, all you need is a U-interface.

TAs, like modems, can be internal or external to your PC. Internal TAs avoid your PC's serial port bottleneck, at 112 kbps, enabling 128 kbps-plus transfer rates. Internal TAs are also generally better integrated with other devices inside your computer than are external TAs.

A TA's compression and bonding standards, proprietary in nature, have to be adhered to at both ends. Make sure that your equipment matches what is installed at your provider's end.

NT1s and NT2s (Network Termination)

NT1s or NT2s attach to an ISDN line at the point where you'd like to split off TAs and other devices for separate ISDN uses.

An NT1 is used for BRI service, usually SOHO and residential applications. Each of up to eight devices connected to an NT1 can have a different ten-digit phone number. NT2s are used for 1.544 Mbps PRI service.

NTs are responsible for performance, monitoring, power transfer, and multiplexing of the ISDN channels. You need an NT that supports multiple S/T interface connections to connect multiple ISDN devices (phones and TAs) to the same ISDN line.

NTs also provide ISDN line power and, therefore, need a source power to operate. This means that if the NT loses power, the ISDN line will not be usable, and it's your problem. It's a good idea to have POTS (Plain Old Telephone Service) lines or power protection equipment for backup.

When buying an NT, find out how it is programmed. The ones that use touch-tone signaling are the simplest to use. Check to see if the provider will provide an NT unit free. They sometimes will, as a promo, or if you ask them.

Operating System Software for ISDN

You will also need the appropriate software so that your communications programs will work with your ISDN line.

Your operating system needs to support ISDN hardware. This support lets applications talk to the ISDN adapter and integrates the ISDN adapter with other parts of the operating system. ISDN adapters usually include a software device driver to integrate with your PC's operating system.

The most common way to connect two computers over ISDN is through the Point-to-Point Protocol (PPP). This is the same protocol usually used to access the Internet. Make sure your Internet Service Provider or remote LAN access server supports PPP, too. If your Internet account is a "SLIP" (Serial Line Internet Protocol) account, you will need to have it changed to PPP.

ISDN Phones

ISDN phones are expensive, but prices are coming down. There are ISDN Centrex operator consoles, too. You'll need EKTS (ISDN Centrex) service from your local phone company, but Caller ID and enhanced call processing features give ISDN phones a general edge over POTS service.

ISDN SOHO Systems

You can set up a flexible, small ISDN SOHO phone system in your small office, home office, or branch office using ISDN BRI service. You'll get multiple telephone lines, directory numbers, and call appearances with ISDN Call Waiting. With these you can have two separate, simultaneous live connections, and eight total connected devices.

ISDN PBXs

ISDN PBXs take ISDN circuits. You can get better ANI/DNIS to your ACD or phone system from an ISDN PRI trunk for your 800-number. Often, station sets will have data ports too. Do credit authorizations or connect to Internet/Intranet servers.

If you're considering ISDN in order to use more than one phone, a fax machine and/or a telephone answering machine, you will need some form of adapter to convert between the digital ISDN signals and the analog devices. All of these adapters could be more expensive to buy than a small ISDN PBX.

A typical, small ISDN PBX set-up has four analog extensions and an internal or external communications bus. Internal buses add quite a bit to the price of the ISDN PBX, but if you need communications between your ISDN phones or PCs, those calls will be handled better than a system that has an external communications bus.

An external bus means that calls between devices on the same system will go through the single main connection.

An internal or external bus have up to eight devices connected to it. Each port has its own internal extension number and can be dialed from anything else connected to the ISSN PBX.

A useful feature of ISDN PBX is off-premises call forwarding, although it requires that another B channel be free when an incoming call is to be forwarded out again. Another useful (optional) feature of most ISDN PBXs allows you to communicate with callers at your door and to operate a remote release of the electric door.

ISDN Video Conferencing

Some video conferencing systems are sold ISDN-ready; just add your ISDN TA for 20 or 30 frames per second video feeds. You'll need a TA, and some H.320-compliant videoconferencing software, plus a camera. Or you can buy a turnkey package that integrates all three.

ISDN PBXs (IPBXs) support video conferencing switch video calls, providing video on-hold, multi-point conferences, hunt-grouping, and voice PBX integration.

ISDN speed video looks like real video; you can actually see and understand the person you're conferencing with.

X.25 over the ISDN D Channel

If your business needs to exchange a small amount of packet data very quickly over a network (yes/no, authorized/not authorized), you can use X.25 networking on ISDN D-channel. ISDN's quick call completion (just 4 to 6 seconds for ISDN vs 30 seconds for analog call completion) helps you serve your customers faster.

Point of sale isn't the only app for X.25 packet data. Service and data inquiries from health care to scratch lottery ticket verification can use packet data. Did I win? Am I covered? Banks and others in the cut-throat data-providing business can gain significant advantages over competitors by offering their customers faster access.

ISDN Line Surge And Power Protectors

ISDN lines, like any twisted copper pair coming into your office/home may not be properly grounded or protected from line spikes. An ISDN line surge protector will keep your equipment from frying.

Power protection is important for ISDN circuits because they are powered by the user, not the telco. If your AC power goes out, you lose your ISDN line(s). ISDN surge/power protectors work like a combination NT1 and Uninterruptable Power Supply (UPS).

ISDN Routers

ISDN Routers are smarter than modems: they dial on demand. They open your ISDN line and dial only when your computer's software requests it, or when your computer receives an IP (Internet) or IPX (Novell network) packet.

ISDN Routers also support multiple simultaneous client access to a network (vs single-client TA-like access). Because ISDN networking devices dial, they make great, viable and less expensive alternatives to leased lines for low-bandwidth data applications. As a bonus, they often give you analog phone/fax management ports in addition to data networking functions.

Serial Port Accelerators

Simple math, here. ISDN goes at 128 kbps over two BONDed B-channels on a BRI line. The serial port on your average desktop PC does only 112 kbps. Put the two together for applications like data networking, videoconferencing and Internet surfing, and you lose bandwidth.

While 16 kbps doesn't seem like much, consider that in remote networking applications especially, ISDN supports effective data transfer rates of up to 512 kbps (achieved by compression, when two boxes that use the same scheme talk to each other) but would bottleneck back down to 112 kbps as soon as you add your PC to the loop. For very little cash indeed you can get back up to speed-and maybe avoid a few IRQ (Interrupt requests, usually from your hardware) or COM port problems by freeing up your PCs built-in COM ports.

ISDN Pricing

ISDN pricing generally has three components

Installation Charge a one time charge to have the ISDN service installed. Part of this charge may be waived if you commit to keeping your ISDN line for a period of time.

Recurring Monthly Charge Depending on the location of the providing carrier, ISDN BRI monthly service charges generally range from $20 to $60. PRI T-1s run from about $400 to over $1000 a month.

Usage charges are associated with use of the ISDN line. Typically it is not more than a couple of cents per minute. The monthly charge may include a certain number of hours of free usage each month. Some packages have no usage charges at all, or may waive charges for use during evenings and weekends.

Be careful. Sometimes ISDN usage charges can double the cost of a call compared to a standard analog POTS (Plain Old Telephone Service) line. It's important to understand how your provider will be billing you.

How to Buy and Install ISDN Service

Your telephone company will perform what is known as line qualification to determine whether your existing wiring will support ISDN.

In cases where the telephone company does not have the right equipment in the local central office that serves you, they can use "line extension" technology to serve you from another exchange. The use of line extension technology may significantly increase the cost of your ISDN service.

For starters, recognize that ISDN equipment is specialized and that no single ISDN product is the ultimate in all three applications: data, voice, and video.

Ask your network administrator or ISP (Internet service provider) which ISDN equipment they support and stick to that list. Even an expert can have a lot of trouble getting an unsupported product to work.

After you've bought the equipment, you're ready to order ISDN service. You'll need a detailed ISDN line configuration, generally supplied in the product manual, from your equipment vendor.

Most data and video configurations are pretty simple. Sophisticated voice applications are the most difficult.

Order your B-channels so that they support both voice and data, especially if there is no extra cost. This flexibility could be important in the future, even if you're only going to use the service for voice at first.

If you are not confident that your order has been received correctly, call back and talk to a supervisor. If the telco doesn't take the order correctly, the ISDN line won't work with your system. When the order-acknowledgment form arrives, check it carefully and be absolutely sure it includes the phone numbers and Service Provider IDs (SPIDs).

A telco installer will come out to install the ISDN line. Most ISDN products have eight-pin connectors and don't work in a six-pin wall jack. The installer should make voice and data calls and run a bit-error-rate test.

Set up the equipment and make a test ISDN call. Make a voice call if your equipment supports it. A voice call is not a test for data communications, but it will show whether the ISDN line is working.

ISDN Configuration: Switch Type and SPIDs

In addition to the configuration the telephone company must do at their end of your ISDN line, there is also some configuration you must do at your end. You need to know three pieces of information supplied by the telephone company to make your ISDN service work.

The Switch Type

Most ISDN hardware adapters need to know what type of switch they are connected to. The switch type simply refers to the type of switch and its software revision level that the telephone company uses to supply ISDN service. There are only a few types of switches in the world and usually just one in countries other than the United States.

Your Phone Number(s)

The second type of information is your phone number or numbers. In some cases, each B channel on an ISDN line has its own number, while in other cases both B channels share a single phone number. Your telephone company will tell you how many numbers your ISDN line will have. Separate numbers may be useful if you plan to take incoming calls on your ISDN line.

Service Profile Identifier (SPID)

The last type of information is the Service Profile Identifier (SPID). The SPID usually consists of the phone number with some additional digits added to the beginning and end. The SPID helps the switch understand what kind of equipment is attached to the line, and if there are multiple devices attached, helps route calls to the appropriate device on the line.

The Future of ISDN

Alternatives to ISDN exist, but ISDN is (finally) here.

As ISDN demand grows, equipment makers will develop a single hybrid product. This will combine NT1, TA and TE1 functions, making the transition from analog to digital easy. You'll be able to make and receive different types of calls with the same equipment.

Consider buying an ISDN router rather than a terminal adapter for your data applications. Routers can cost twice as much but are faster and can be configured remotely, which means you or anyone else can troubleshoot from anywhere else.

Frame Relay

Frame relay is a digital voice/data delivery service sent over T-1 circuits running at speeds ranging from 56 Kbps to 1.5 Mbps (full T-1 speed).

Frame relay is an established packet-switched technology, widely used and well understood by telcos. Almost every major long distance and local telephone company offers frame relay, and since the Telecom Act of 1996, so do CLECs. There are tons of applications and products.

Frame relay lets you transmit data, voice or video using variable-length packets known as frames. Frame relay is engineered for building virtual private networks (VPNs).

What is a VPN?

VPNs have been offered by long distance carriers for years, what's different from one to the next is the way the data is sent, and which carrier is sending it. VPNs look like dedicated, point-to-point circuits connecting your offices or branches, except that VPNs aren't dedicated, or point-to-point. You purchase "here to there" connections and a guaranteed data rate, and every time you make a call the carrier's network switches the calls over the best path for them (based on the carrier's traffic levels, commitments and costs) at that particular time.

Frame Relay Beginnings

Packet-based data transport services have been available since the 1970s.

The private leased-line networks used before packet switching was available were point to point, dedicated circuits. Leased-line networks tended to be over-trunked because providing good service (minimal blocked calls) meant you needed to buy enough physical connections to handle peak demand. Any other time, the circuits are paid for but not used to capacity. Packet switching improves network efficiency by reducing over-trunking, because every connection is made on demand.

Frame relay's predecessor, X.25, was designed to run over existing copper cable. It used a big part of each packet to carry error-checking routines, which made its upper-limit bandwidth 56 Kbps.

The benefit of error correction to a network decreases exponentially with reduced error rates. When the private line physical facilities interconnecting the packet switching systems improved in the 1980s, error rates dropped to very low levels, and the extra cost (in terms of bandwidth or speed) for correcting each additional error became prohibitively high.

The frame relay protocol doesn't include the error correction overhead that is required by X.25, so it can transport more information in the same amount of time.

WilTel now part of LDDS, introduced frame relay services in 1991. Frame relay really gained popularity in the mid-90s when Internet service providers migrated their private line connections between their routers to frame relay.

How Frame Relay Works

Packets are made up of a header, some signaling bits, and an information field, or payload. Each frame relay header is 2 bytes in length and references a permanent or switched virtual circuit (PVC or SVC). The header also contains congestion and status signals from the network to the user. The information field may include other protocols within it, such as an X.25, IP or SDLC (SNA) packets.

Permanent Virtual Circuits (PVCs) Most frame relay networks are built using preprogrammed routes, PVCs, through the network from source to destination.

PVCs make switched frame relay services look like dedicated, always-connected private lines. The packet-switching network can use any of a number of paths for your call and some path between two points will always be available.

Switched virtual circuits (SVCs) provide connections on demand, but aren't physical connections. They work like dialup services. Switched virtual circuits (SVC) are easy to configure, operate and manage, and further reduce latency for voice. SVCs improve latency by providing cut-through switching across the network. For voice applications, the network acts as a virtual tandem PBX providing one-hop switching between each pair of PBXs.

Multiplexing Techniques

Some equipment vendors offering voice FRADs (Frame Relay Access Devices) use different bandwidth optimization multiplexing techniques such as Logical Link Multiplexing and Subchannel Multiplexing. Logical Link Multiplexing allows voice and data frames to share the same PVC (Permanent Virtual Circuit). This can provide savings on carrier PVC charges and it increases the utilization of the PVC.

Subchannel Multiplexing is a technique used to combine multiple voice conversations within the same frame. By allowing multiple voice payloads to be sent in a single frame, packet overhead is reduced and performance on low speed links is efficiently increased.

Frame Relay Access Devices (FRADs)

A frame relay access device (FRAD) sends packets of data to any virtual circuit in your VPN (Virtual Private Network) by writing the address of the proper circuit into the DLCI (Data Link Connection Identifier) field of each packet's header. A single, voice-capable FRAD (programmed with a custom dialing plan) can establish talk channels with compatible devices anywhere in your VPN where virtual circuits have been made available.

In addition to providing basic services such as encapsulating data traffic for transport over the frame relay network, voice capable FRADs (Frame Relay Access Device) may sometimes provide connectivity between PBXs and other voice equipment.

Since frame relay evolved from X.25, the first generation frame relay switches were modified X.25 switches. These switches tend to be bound by the storage, processing and forwarding requirements of X.25. As a result, first generation switches may add significant latency, especially during congested periods. On the other hand, the discard rate may be reduced since frames can be stored until a circuit is available.

At the other end of the spectrum are the combination Frame and ATM switch platforms. These systems are engineered for maximum throughput at the switch, and cause little or no latency. However, the switch platforms forward the packets without regard for congestion. If the FRAD's communications bus or processor card is incapable of forwarding packets fast enough, it will start to discard them. Eventually, Quality of Service (QoS) will suffer.

First-generation FRADs were data-only devices. Next came what are called internetworking FRADs, which integrated IP (Internet Protocol) routing functionality. Next came voice FRADs, which integrated voice support. All three have these common networking attributes: they use frame relay permanent virtual circuits or PVCs, operate at a single class of service for voice and data, and require a FRAD at both ends. FRADs provide fragmentation of data frames on an end-to-end basis to prevent latency when switching voice calls.

Voice-enabled frame relay equipment

Branch office FRADs are the, smallest, providing a limited number of premises-side connections and limited throughput. For data input on the premises side, they generally provide programmable serial ports, which can accept output from a router, or RS-232 to accept output from terminals.

For voice/fax connectivity, FRADs provide a limited number of voice ports (two or four) and a range of electrical options. (FXS/FXO, ground start/loop start, E&M, etc.) On the network side, most FRADs come, or can be equipped with a V.35 serial port for hooking to a DSU (Digital Service Unit).

Regional office FRADs are larger. They're usually card cages in to which a wide variety of option cards can be installed-additional data-input ports, additional analog voice ports, digital voice ports, direct Ethernet connections, router cards, DSU cards, etc.

Central office FRADs offer the most flexibility, greater throughput and emphasize direct T-1 connections to a large phone system. CO FRADs serve as hubs for both data and voice traffic, reducing the number of (virtual) circuits required to support a VPN.

CODEC (COder/DECoder)

While relatively easy to set up, frame relay isn't the friendliest possible network for transmitting voice. Frame relay is non-isochronous, accepts packets of widely-varying sizes which reduces header-to-packet overhead ratios, but is diffi-

317

cult to manage because mixed voice and data use the same communications links.

Top of the line Frame Relay equipment offers a range of CODEC (COder/DECoder) options that compress and transmit the primary voice signal at different bandwidths.

At this point, most manufacturers are going with CELP/ECELP (Code-Excited, Linear Predictive, and EnhancedCLEP) CODECs at 8 Kbps per voice channel or 16 Kbps per voice channel, most of which follow a subset of the proposed G.729 standard; and they also usually provide "proprietary" CODECs for narrower bandwidth applications.

You can program the bandwidth you need, per channel and establish fallback selections for times when contention for bandwidth is high. Remember, though, that all of these systems are essentially proprietary. You still need to buy all your FR gear from one manufacturer in order to guarantee compatibility.

Voice Over Frame Relay (VoFR)

Frame relay virtual circuits can also carry voice and fax — the same way point-to-point leased digital lines can — when equipped with CSU/DSUs, muxes, etc. Your interoffice voice and fax traffic over your private frame relay network doesn't have incremental or per-call charges.

Because voice over frame relay (VoFR) employs profound voice compression, it can squeeze both data and voice simultaneously through very narrow bottlenecks without severely compromising performance. And unlike channelized approaches, when you're not transmitting voice, all bandwidth becomes available for data. Many users discover that they can add a moderate amount of voice to an existing frame relay setup without suffering any decrease in apparent throughput, even on the narrowest connections in their network.

Even when existing bandwidth is efficiently utilized, some network managers might find that the incremental cost for the additional frame relay network bandwidth needed for voice transport is more cost effective than some of the standard voice services offered by local and long distance carriers.

There are some problems with VoFR. These include some loss of voice quality compared to call traffic over the PSTN, because VoFR uses voice compression, and the loss of consolidated voice billing and invoice itemization, end user charge back capabilities, and other advanced features such as ID and accounting codes. Other problems with VoFR include the lack of equipment interoperability between customer premises equipment vendors, and the lack of standards defining the acceptable levels of quality for voice transport over a carrier's frame relay network.

Specific VoFR service guarantees and advanced carrier troubleshooting capability are lacking because the implementation of VoFR occurs in the equipment at the end users premises, and control is therefore outside of the carrier's frame relay network.

Committed Information Rate (CIR) and Excess Information Rate (EIR)

Perfect QoS management anticipates all possible congestion points. Your data can be slowed by your own equipment, by your provider's equipment, and by the provider's available bandwidth.

Frame relay services offer a minimum Quality of Service (QoS) through two bandwidth parameters, the Committed Information Rate (CIR) and the Excess Information Rate (EIR).

Frame relay virtual circuits aren't leased in terms of an absolute bandwidth commitment. Instead, they're leased in terms of a nominal burst rate and a "Committed Information Rate."

The burst rate is the top speed (largest bandwidth) your network will need for everyday operations. The CIR is a lower data rate that the network provider is committed to provide at all times. The CIR establishes the minimum amount of bandwidth that will be available at any time.

By changing the CIR, you guarantee yourself enough bandwidth so that your applications won't choke when the network gets congested.

If, for some reason, your network requires more bandwidth than the CIR you agreed to, insurance is available.

The EIR allows bursting above your CIR, if your provider has bandwidth available and you want to pay for data overdraft protection. EIR guarantees make frame relay attractive for data backup performed at night when the frame relay backbone networks are charging off-peak rates.

You can negotiate cheaper rates from your provider using a relatively low CIR, so your flat monthly fees are low, and with additional EIR guarantees, actual network access rates will probably be higher than the CIR. You'll get more bandwidth for less money.

How to Buy Frame Relay Services

First, call your carrier and order frame relay service at connection bandwidths appropriate to overall bandwidth requirements at each of your locations. The carrier brings in this service on 56 Kbps pairs or over fractional or whole T-1s to your premise. You can probably get away with less bandwidth than you think.

At the same time, you determine the maximum burst bandwidth requirements and CIR's of the permanent virtual circuits that link your locations. If you want fastest throughput, you can use an "every to every" design, in which each location terminates a virtual circuit to every other. Or you can save on recurring charges for the virtual circuits by linking all remote locations to a frame relay "hub" at a central location that can act as a switch, connecting virtual circuits to make connections between branch offices through the central site.

Buying frame relay equipment isn't difficult, either. Voice/fax-capable frame relay access devices (FRADs) provide premises-side interfaces for data equipment and they connect to your office phone system via loop start or ground start trunk ports or, for the largest systems, T-1 trunk connections. Basically, you order equipment with the interfaces required to hook up to the equipment that you already own.

On the voice side, branch offices equipped with key systems will want to dedicate one or two trunks to interoffice use, so people can press a dedicated key to reach the Frame Relay interoffice network. PBX owners may want to adopt a "dial 8" plan, and dedicate several trunk connections to their voice FRAD.

Finally, you program the FRADs with your dialing-plan and voice-handling parameters. You can make them act as ringdown circuits via virtual circuit to different branches and extensions or build a multi-digit dial-through plan.

Present and Near Future Frame Relay

ATM/cell relay is gradually replacing frame relay.

ISDN and frame relay overlap within the data communications market. Interoperation between ATM and frame relay is another natural overlap. The Frame Relay Forum and the ATM Forum have defined standards for interworking these two protocols.

Over time, frame relay will probably be used as a multimedia access channel to higher-bandwidth, QoS-rich ATM networks.

Asynchronous Transfer Mode (ATM)

Asynchronous Transfer Mode (ATM) is a digital transport technology that can simultaneously transmit voice, data, images and video. Also known as Cell Relay or Broadband ISDN, ATM operates at speeds from 155 Mbps (million bits per second) to 622 Mbps. It is classified as an asynchronous transport technology because the cells carrying data are not required to be periodic, or timed.

ATM grew to fill a need for a worldwide standard to send digital information, regardless of the "end-system" or type of information sent. ATM was designed to provide large amounts of bandwidth economically and on-demand. It is predicted to be the next worldwide digital network transport technology for large public and private carrier networks.

Separate networks have traditionally carried voice, data and video information primarily because different types of traffic have different characteristics and bandwidth requirements. Data traffic tends to be "bursty"; there is no need to communicate for an extended period of time and then all of a sudden there is a need to communicate large quantities of information as fast as possible. Voice and video, on the other hand, tend to require regular amounts of bandwidth, but are very sensitive to when and in what order the information arrives.

ATM's short cell length minimizes end-to-end processing delays and is works better for voice, video, and other delay-sensitive protocols. OSI-like adaptation layers (more later) accommodate a wide variety of machines with different physical connections running different type of communications.

ATM-transmitted information is segmented into standard, 53 byte pieces, called cells, and then sent and re-assembled at the destination. Fixed length cells allow information to be transported in a predictable manner. This predictability lets the network successfully schedule and deliver different traffic types over the same network.

ATM cells are transmitted serially in strict numeric sequence over either broadband or narrowband transmission facilities.

Why ATM?

ATM is a switched technology. Switched calls mean dedicated bandwidth for every connection, higher aggregate bandwidth, well-defined connection procedures and flexible access speeds.

ATM is more efficient than Time Division Multiplexing (TDM), the current digital transport technology used for voice, because it dynamically shares available bandwidth among multiple logical connections instead of slicing the bandwidth into fixed-size channels dedicated to particular connections. When a user doesn't need access to a network connection, the bandwidth is available for use by another who does.

ATM is fast, operating at speeds into the gigabits-per-second range. Fixed-size cells with a simple and consistent format enable switching functions to be performed in hardware, and also result in reduced queuing delays.

ATM is a multimedia transport method. Virtually any type of information can be converted into cells for transmission over an ATM network.

In the ATM transmission of a steady stream of small fixed-size cells, information contained in the cell headers affiliates each cell in an ATM stream to a particular logical connection. Connections needing more bandwidth are simply allocated more cells, and when a connection is idle, no cells are generated, and no bandwidth is consumed.

Voice and video are very sensitive to delay, and the delay cannot be allowed to vary much. Data, on the other hand, is not nearly as delay sensitive as voice and video, but is very sensitive to loss. ATM gives voice and video traffic priority with fixed delay, while simultaneously making sure that data traffic has low loss. The way ATM accomplishes this is with virtual connections.

Permanent Virtual Circuits (PVCs) and Switched Virtual Circuits (SVCs)

ATM has two virtual circuit communications services. Switched Virtual Circuits (SVCs) establish short-term connections that require call setup and teardown. Permanent Virtual Circuits (PVCs) are similar to dedicated private lines because the connection is set up on a permanent basis.

ATM virtual connections can operate at a constant bit rate (CBR) for voice and video traffic, and at a variable bit rate (VBR) for data traffic. Each virtual connection has its own set of parameters: Minimum Cell Rate (MCR), Sustained Cell Rate (SCR), and Peak Cell Rate (PCR), that help the ATM network to determine the bandwidth, relative priority, and Quality of Service (QoS) that should be assigned to the call.

How ATM Works

The sender negotiates a "requested path" with the ATM network server, specifying the type, speed and destination of the call, which determine the end-to-end quality of service that the network will assign to the call.

Time-sensitive transmissions like voice and video are given a higher required QoS (Quality of Service) and bandwidth as needed until an upper bandwidth limit is reached.

The ATM switch takes the user's information, breaks it up into cells, and then the switch multiplexes the cells together into a single bit stream and transmits it across the digital link.

Each ATM cell is has two main sections, the header and the payload. The payload (48 bytes) is the part that carries the actual information-voice or data or video. The header (5 bytes) contains the address and queuing information.

At any given time, some ATM virtual circuits will be transmitting a burst of data over the shared trunk, while others are quiet. ATM only allocates bandwidth to a connection when there is data to send.

When the cells reach their destination, ATM adaptation processes unpack the cells and convert them back into their native formats — frames for data processors, and constant delay bit streams for voice and video devices. Finally, the data is delivered in sequence to the appropriate device.

ATM Standards and Regulatory Bodies

Two types of standards bodies are actively involved in B-ISDN and ATM: formal standards bodies and industry forums. Industry forums are independent groups of industry experts, vendors, and users concerned about interoperability issues.

National and international interoperability or standardization, is a prerequisite to success in the data communications marketplace. ATM has been chosen as the standard transfer mode technology for the Broadband Integrated Digital Network (B-ISDN), and it is within B-ISDN that ATM is standardized.

The four industry forums actively specifying B-ISDN/ATM are the ATM Forum, the Frame Relay Forum, the Internet Engineering Task Force (IETF) and the SMDS Interest Group (SIG).

The formal standards organizations that are active in specifying B-ISDN and ATM are the International Telecommunications Union (ITU) -Telecommunications Standardization Sector (formerly called CCITT), The American National Standards Institute (ANSI) in the United States, and the European Telecommunications Standards Institute (ETSI) in Europe.

The B-ISDN Protocol Reference Model

B-ISDN describes ATM functionality using a layered reference model similar to the Open Systems Interconnection (OSI) 7-layer architecture. The B-ISDN Protocol Reference model redefines the lower three layers as the Physical Layer, the ATM Layer, and the ATM Adaptation Layer (AAL).

> See the *Data Communications and Digital Voice* chapter at the beginning of this section for a description of the OSI Model.

The higher-order layers are software intensive, and are associated with specific user applications residing on the user devices that are served by the ATM layers. The lower layers of the B-ISDN Protocol Reference model are more hardware-intensive.

The Physical Layer

The Physical layer of the ATM model defines the electrical characteristics and network interfaces of the transmission. This layer defines the physical interface, transmission rates, and how the ATM cells are converted into a line signal.

There are two sublayers to the Physical Layer that describe the physical transmission medium (called the Physical Medium Dependent (PMD) sublayer) and the data extraction function (the Transmission Convergence (TC) sublayer).

The physical medium sublayer controls transmission and receipt of a continuous flow of bits with the associated timing information. It synchronizes the transmission and reception of the data for functions that are dependent upon the physical medium, like the transmission rate and the physical characteristics of the connectors.

Some existing physical medium standards for ATM cells include SONET (Synchronous Digital Network or SDH), DS-3/E3, 100 Mbps FDDI (fiber distributed data interface) and 155 Mbps Fiber Channel.

The transmission convergence sublayer extracts the information content from the physical layer data transmission. It performs Header Error Correction (HEC) generation and checking, extracts cells from the incoming bit stream, and processes idle cells.

The ATM Layer

The ATM Layer is the layer that deals with the structure of an ATM cell. Each ATM cell consists of a 5-byte header and a 48-byte payload. The header contains the ATM cell address and other switching information. The payload contains the user data being transported over the network.

The standards committees have defined two types of ATM cell headers: the User-to-Network Interface (UNI) and the Network-to-Network Interface (NNI). The UNI is a native-mode ATM service interface to the WAN (Wide Area Netework). Specifically, the ATM UNI defines an interface between cell-based customer premises equipment (CPE), such as ATM hubs and routers, and the ATM WAN.

The NNI defines an interface between the nodes in the network (the switches) and interfaces between networks. The NNI can be used between a private ATM network and a service provider's public ATM network.

Both types of cell headers, the UNI (the I stands for Identifier) and the NNI, identify virtual paths (VPIs) and virtual channels (VCIs) as routing and switching identifiers for ATM cells. The VPI identifies the path or route to be taken by the ATM cell, while the VCI identifies the circuit or connection number on that path. The VPI and VCI are translated at each ATM switch, and are unique for a single physical link.

The ATM Adaptation Layer (AAL)

The Adaptation layer divides all types of data into the 48 byte payload that will makes up an ATM cell.

The function of the adaptation layer (AAL) is to accommodate data from various sources with differing characteristics. It adapts the services provided by the ATM layer to those that are required by the higher user layers (such as circuit emulation, video, audio and frame relay).

The AAL receives the data from the various sources or applications and converts it to 48 byte segments that will fit into the payload of an ATM cell.

The adaptation layer defines the basic principles of sublayering. It describes the service attributes of each layer in terms of constant or variable bit rate, whether the transmission requires timing, and whether the service is connection-oriented or connectionless.

AAL1 is used for a constant bit rate, connection-oriented service that requires timing (synchronous), like voice and video.

AAL2 is used for a variable bit rate, connection-oriented service that requires timing (synchronous), like compressed voice and video.

AAL3/4 is used for a variable bit rate, connectionless service that doesn't require timing (asynchronous), like SMDS and LANs.

AAL5 is used for a variable bit rate, connection-oriented service that doesn't require timing transfer (asynchronous), like X.25 and Frame Relay.

Any AAL is made up of two sublayers, the Convergence Sublayer (CS) and the Segmentation and Reassembly Sublayer (SAR). The Convergence sublayer receives the data from various applications and packetizes it into variable length data packets called Convergence Sublayer Protocol Data Units (CS-PDUs).

The Segmentation and Reassembly sublayer receives the CS-PDUs and segments them into one or more 48 byte packets that map directly into the 48 byte payload of the ATM cell transmitted at the physical layer.

ATM Traffic Management

ITU-T Recommendation I.371 specifies how user traffic can be policed. There is an agreement, a contract, between the user and the network regarding the Quality of Service (QoS), and the traffic parameters or descriptors that regulate cell flow.

The traffic descriptors depend upon the particular class of service, and may include a sustained cell rate (SCR), a minimum cell rate (MCR), a peak cell rate (PCR), and/or a burst tolerance (BT). Because ATM can carry delay-sensitive voice or video traffic together with less delay-sensitive data traffic, the switch must not only ensure a guaranteed amount of bandwidth, but also an acceptable level of delay and delay variation for voice and video traffic.

The ATM Forum addresses the issues of traffic shaping, which regulates traffic before it gets into the network, and congestion control, which regulates traffic within the ATM network. Traffic shaping, or connection admission control algorithms, ensures that traffic on a given virtual connection conforms to the Quality of Service agreement at its source and minimizes the possibility of cells being discarded inside the network.

Internet Protocol (IP) over ATM

The success of Asynchronous Transfer Mode (ATM) depends upon its ability to transport legacy data traffic. Since the explosion of the Internet, it is very important that internet protocol-based data interoperates with ATM. The problems of running IP data over ATM networks are due to the differences between them:

ATM is a switched service; a connection has to be established between the sender and the receiver before data can be sent. Once the connection is set up, all data between them is sent along the predefined connection path.

IP is a packet-based connectionless protocol. Each IP packet is forwarded by routers independently on a hop-by-hop basis.

To send IP traffic over an ATM network, either a new connection is established on demand between the sender and receiver, or the data is sent through a preconfigured, dedicated connection or connections.

When the amount of data to be transferred is small, the cost of setting up and tearing down a connection is not justified. Preconfigured path(s) may not be the optimal paths at the time of transmission, and may become overwhelmed by the volume of data being transferred.

Quality of Service (QoS) is an important concept in ATM networks. It includes the parameters of the bandwidth and delay requirements of a connection. Such requirements are included in the signaling messages used to establish a connection. Current IP (IPv4) has no QoS concept and each packet is forwarded on a best effort basis by the routers.

To take advantage of the QoS guarantees of the ATM networks, the IP protocol needs to be modified to include the QoS requirements of the transmission before legacy IP traffic can be supported by ATM. Treating ATM as yet another LAN technology, classical IP over ATM is easy to implement. At this writing, IP data sent over ATM networks has to travel through a router even when both parties are directly connected to the ATM network.

ATM Hurdles

Management Issues

Network management implies the ability to monitor and control every element in an ATM network such as local customer premises equipment, local ATM switches, wide area network interfaces, and wide area services. End-to-end management issues have not been resolved for existing network technologies, and so addressing them for ATM has not progressed very far.

The ATM Forum has defined the Interim Local Management Interface (ILMI) that is based upon standard SNMP protocol. Until ATM network management standards are fully defined, the ILMI provides early users of ATM with an interim set of interface management functions. These guidelines describe how to manage status, configuration, and control parameters that are present on a private or public UNI.

ATM Supply and Demand

Carriers and providers are unwilling to make the necessary investment in ATM without pre-existing demand, and corporations are unwilling to invest in new broadband applications and access equipment without pre-existing ubiquitous and economical services.

Carrier networks are optimized for voice communications, because that has been the type of traffic type most often carried over long distances.

Local area network structures are optimized for data, because data applications running on LANs justified user investments in LAN technology.

ATM, which includes both voice and data traffic types, can't be delivered over either the existing local data network infrastructure or the existing carrier network infrastructure. Any new service begins with no network and no compatible CPE (Customer Premises Equipment).

There's a demand for scaleable ATM-based switching platforms that support multiple access interfaces at narrowband (lower) speeds. ATM is defined at speeds of 45 Mbps, 100 Mbps, 155 Mbps, and up to 622 Mbps by the ITU-T standards. Most end users can't justify the high bandwidths defined for ATM connections-56 kbps WAN connections still outsell 1.544Mbps T-1/E-1 connections by at least 3:1.

Present and Future ATM

To achieve high speed LAN/WAN connectivity, some users are building hybrid networks that are a combination of private data networks and public WAN services. Other companies are using public carrier WANs that use ATM.

Service providers deploying ATM networks with frame relay interfaces gain a flexible network infrastructure that is cost-justified by the exploding demand for frame relay services. Later, they can begin to offer native ATM interfaces as the demand for them grows.

Digital Subscriber Line Technologies (xDSL)

xDSL is fast digital communications, for voice and data, over your existing local copper wires.

DSL stands for Digital Subscriber Line (or Loop). The communications technologies collectively called xDSL, are variations on a theme that enables fast (high bandwidth) digital communication of data, and at the same time allows an analog signal for voice.

ISDN and T-1 circuits are powered at your end, by your equipment. If your power goes out, no data, and no dial tone, either. It's common sense (and common practice) to buy extra analog circuits for backup if you use these technologies.

You don't have to spend extra for a backup analog line with xDSL. If your xDSL hardware fails or if you have a power outage, you'll still get dial tone.

xDSL modems send data on top of POTS (Plain Old Telephone Service) service using frequencies higher than the 3.4 kHz needed for analog voice. The telephone company filters out these high frequencies at the CO, and then the voice calls are handed off to the PSTN (Public Switched Telephone Network) to their destinations. Voice signals don't have to go through costly digital-to-analog conversions at the CO switches.

xDSL is cheaper to buy and install because it uses two wires, not four, per connection. Companies that provide xDSL can serve twice as many people over the same copper plant, and end users only have to rent one pair of wires, not two.

The serial data modulation scheme used for xDSL means simpler circuitry and cheaper hardware, and xDSL's hardware has integrated router capabilities.

Unlike ISDN, xDSL is designed to degrade gracefully in response to conditions on your phone line. ISDN is very intolerant of noisy lines (it either works correctly or not at all). xDSL lines will slow down as the line becomes noisier, but conditions have to become very poor for the circuit to fail entirely.

xDSL Basics

xDSL modem technologies have bandwidth and distance trade-offs. The longer the length of the copper wires between the Central Office and end user, the lower the bandwidth. As distance decreases toward the telephone company office, the data rate increases.

xDSL carries several different signals on the same cable simultaneously. The total bandwidth is divided into separate frequency ranges, called carriers. xDSL carriers all carry different parts of the same data transmission at the same time and reassemble them at the remote end.

xDSL is point-to-point communication (from one CO to one subscriber or between two subscriber points over a leased copper line), with no Central Office hops or repeaters in between. Each computer on an xDSL network is addressed using a valid IP address from either the public or a private address range. If you own your Internet address range, your address won't change when you use xDSL.

ADSL (Asymmetric Digital Subscriber Line) and VDSL (Very High-Speed Digital Subscriber Line) have an asymmetric data flow, meaning that more data is sent in one direction than the other. These technologies are ideal for applications like Internet access, remote client/server links and video because there is more bandwidth used in the downstream (receiving to the customer) direction than upstream to the server.

xDSL networks have five parts: DSLAMs (Digital Subscriber Line Access Multiplexers) at the network end, xDSL modems, POTS splitters, CSU/DSUs (Customer Service Unit/Digital Service Unit), and Ethernet NICs (network interface cards) in connected PCs at the customer end.

DSLAMs handle incoming xDSL modem calls to the telco Central Office. The simplest DSLAMs provide high-speed input/output of the raw data-stream; deluxe models may integrate protocol converters and additional hardware for connecting to other types of data networks.

At both the premises and the network end POTS splitters provide regular CO phone services and CSU/DSUs provide interfaces for hooking up T-1 phone/data equipment. Most xDSL modems can also function as routers on an Ethernet LAN so multiple PCs can connect through cheap Ethernet NICs (Network Interface Cards).

POTS splitters are optional. However, if you or your local service provider don't use POTS splitters, then the ADSL data service is supplied to you on a second pair of copper wires.

xDSL Applications

Remote connections. You can connect two PBXs to work as one, with an xDSL T-1, or, connect your branch offices to corporate networks. Small offices can direct-connect to an Internet Service Provider (ISP) for very little money. Telecommuters can have phone and data/videoconferencing services delivered over a single copper pair.

Internet access. The bandwidth available over xDSL provides much quicker downloads than analog modems or ISDN BRI Internet. You can download graphics and audio intensive applications plus interactive 3-dimensional applications-something that is not possible using slower access technologies.

Video

Full-motion video is possible with xDSL. Video applications over ADSL use a bandwidth of about 1 MHz, designating the frequencies between 50 kHz and 1 MHz for downloaded video data. The lowest 4 kHz is dedicated to POTS and the 40 kHz in between serves as the upstream signal path.

Pair gain. In so-called "pair gain" applications, xDSL modems squeezes two analog phone connections onto a single pair of wires for maximum connectivity where copper wiring is scarce.

Real-time, at-your-request music. You can connect to your service provider's network over your xDSL line and access a database of available music-on-demand which will then be played back to you, now or at some later time.

Remote learning. A whole classroom of students can view a professor's presentation, interact with the instructor during the lecture, and take tests remotely, all while sitting at a PC nowhere near the classroom.

Medical records. A doctor can get patient records, quickly, from other health care facilities. A patient's medical images could be transmitted to a specialist while the patient's doctor is consulting with the specialist on the phone.

xDSL Technologies

Asymmetric DSL (ADSL)

ADSL (Asymmetric Digital Subscriber Line) has a16 to 640 Kbps upstream bandwidth, a 1.544 Mbps downstream bandwidth and a 18,000 foot range.

Introduced in 1989 by Bellcore Laboratories, ADSL works on the basic premise that most of the data during transmission will go in one direction: from the server to the consumer.

ADSL is called "asymmetric" because most of its two-way or duplex bandwidth is devoted to the downstream direction, sending data to the user. Only a small portion of bandwidth is made available for upstream or user-to-server messages.

This unequal capacity works well for remote computing and Internet surfing because these end users access more information than they generate.

Unless your phone company offers "splitterless" ADSL (see DSL Lite), you will need to install a xDSL modem in your computer. ADSL modems have varying speeds and capabilities. The least expensive allows 1.5 or 2 Mbps downstream and a 16 Kbps duplex channel; others provide rates of 6.1 Mbps and 64 Kbps duplex. Products with downstream rates up to 9 Mbps and duplex rates up to 640 Kbps are available at the top end.

DSL Lite

DSL Lite is a slower version of ADSL that doesn't require splitting of the data and voice signals at the user end; the Central Office splits the data and analog voice components for the user remotely. Remote splitting reduces the maximum data rate to 1.544 Mbps, which is still higher than current modem data rates.

Consumer DSL (CDSL)

CDSL is a trademarked version of xDSL that is slower than ADSL (at 1 Mbps downstream), but, like DSL Lite, has the advantage that a splitter does not need to be installed at the user's end. Rockwell owns the technology and makes a chipset for it. CDSL uses Rockwell's proprietary carrier technology.

High-speed DSL (HDSL)

HDSL has a 1.544 or 2.048 Mbps upstream bandwidth with the same bandwidth downstream and a 15,000 foot range. The main characteristic of HDSL is that it is symmetrical: an equal amount of bandwidth is available in both directions. Since the bandwidth is split equally in both directions, the maximum data rate for HDSL is lower than that for ADSL.

HDSL was the first xDSL technology arising as a solution to the problem of a carrier trying to transmit broadband T1 signals (1.544 Mbps) over long copper loops. Long loops (the copper wires that carry the signals from the nearest CO to the home or business) distort signal quality over distance, so repeaters or amplifiers are installed on copper pairs at prescribed intervals to boost and maintain acceptable signal quality.

Today's T-1 networks have repeaters installed about every 3,000 to 4,000 feet. Installing and maintaining the repeaters is time consuming and expensive, and makes T-1 circuits more expensive for customers that are farther away from Central Offices.

HDSL T-1 transport needs only a single pair for short loops, and can reach to the full 18,000 feet on two pair.

When HDSL runs on two pairs, each set of wires carries a full-duplex signal, using echo cancellation, at half the T-1 rate. A T-carrier uses one pair for each direction at the full 1.5 Mbps. The ILECs use HSDL extensively now for point-to-point T-1.

Symmetrical DSL (SDSL)

SDSL has a 1.544 (United States) or 2.048 (Europe) Mbps upstream bandwidth, the same downstream bandwidth and a 10,000 foot range.

If the speed is the same in both directions, you have SDSL. Vendors that provide SDSL use DMT (Discrete Multi-Tone) encoding. SDSL will be an important future implementation because of its ability to carry T-1 on a single pair.

ISDN (Basic Rate) DSL (IDSL)

IDSL has an upstream bandwidth of 144 Kbs; downstream is the same and it has a15,000 foot range. It is exactly the same as the ISDN basic rate interface (BRI): 2B1Q (which means two "bearer" or data channels and one signaling channel) running at 80,000 symbols per second to carry 160,000 bps.

The phone companies will probably find IDSL easy to accept because they've had 20 years experience with ISDN BRI. Customer equipment vendors like the idea because IDSL uses the same CPE (ISDN terminal adapter, bridge or router) as ISDN; only some software changes are needed to convert existing customers.

Several makers of central office equipment are jumping into the xDSL market using the IDSL format because the ISDN chips they need are readily available and familiar to their technicians

IDSL has the advantage of being viable over longer local loops - greater than 18,000 feet from the central office, and over lines that have repeaters or Digital Loop Carriers (DLC) installed on them.

Medium-speed DSL (MDSL)

Medium-speed means fractional T-1. Any of the xDSL protocols will carry fractional T-1, but 2B1Q (ISDN BRI) is used most because it is simple to put on a single pair of wires.

Rate-Adaptive DSL (RADSL)

RADSL is a rate-adaptive ADSL technology from Westell whereby the equipment determines the rate at which the signals can be transmitted over each customer loop and adjusts the delivery rate accordingly. Westell's FlexCap2 system uses RADSL to deliver from 640 Kbps to 2.2 Mbps downstream and from 272 Kbps to 1.088 Mbps upstream.

Very-high-speed DSL (VDSL)

VDSL has a 1.6 to 2.3 Mbps upstream bandwidth, a 13 to 52 Mbps downstream bandwidth and a 1,000 to 4,500 foot range.

If you want to go really fast, like 52 Mbps, you need an optical fiber. Or, if the distance is short (up to 1000 feet), you can use VDSL. This available distance is fine for Local Area Networks.

VDSL is a developing technology that's envisioned to emerge somewhat after ADSL is widely deployed and then co-exist with ADSL. The specific transmission technology is not yet determined, but a number of standards organizations are working on it.

x2/DSL

x2/DSL is a modem from 3Com and US Robotics that supports 56 Kbps modem communications now, but will be upgradeable with new software for ADSL when it becomes available in your area. 3Com calls it "the last modem you will ever need."

For the time being, you'll be buying the xDSL equipment your service provider supports, and using it for whatever service your service provider sells, since not all xDSL schemes are interoperable. Find out what your provider is using before you buy your equipment.

xDSL Technologies and Standards

Each end of a xDSL loop has special hardware to encode and decode the data signal and is called an active headend.

Discrete Multi-Tone (DMT) has been established as the ANSI and ETSI standard for ADSL. Carrierless Amplitude and Phase (CAP) modulation, while not the standard, has been widely implemented side-by-side with DMT.

DMT (Discrete MultiTone) divides the 1 MHz spectrum available over a copper local loop into 256 4 KHz channels. It varies the bit densities (how much information is sent at one time) on each of these channels to minimize noise and interference that could impede the data flow over sections of that spectrum. One of DMT's advantages is it maximizes throughput on good channels and minimizes throughput on channels with heavy interference.

CAP relies on a single carrier and uses techniques similar to the Quadrature Amplitude Modulation (QAM) used in V.34 (28.8 Kbps) modems.

Both DMT and CAP use an ADSL modem on each end of a twisted-pair telephone line and create three information channels: a

343

high speed downstream channel at speeds from 1.5 to 8 Mbps, a medium speed duplex channel that runs from 16 to 640 Kbps and a POTS (Plain Old Telephone Service, or analog) channel.

Despite the ratification of the DMT standard, modems utilizing both methods are being tested by carriers to gauge the cost of each ASDL algorithm's deployment cost and performance. Both encoding algorithms have received positive ratings for performance, so if both perform similarly, cost and manageability is likely to be the deciding factor between the two. CAP costs less than DMT, but DMT offers greater flexibility in terms of bandwidth allocation.

The International Telecommunications Union's (ITU) standards development is a two stage process, Determination (technical approval), then Approval (previously termed Decision). The period between Determination and Approval allows manufacturers the time to implement the draft Recommendation, prove proper operation, and identify any necessary technical changes that need to be made to the draft Recommendation.

In October 1998 the six international standards (Recommendations) necessary for the reliable, compatible and maintainable use of ADSL were Determined. Here are the specifics:

G.992.1, ADSL Transceivers

This Recommendation is an ADSL standard for network access at rates up to 6.144 Mbps downstream and 640 Kbps upstream using the Discrete Multitone (DMT) protocol, which is similar to the way modems modulate signals. G.992.1 is modeled after American National Standards Institute (ANSI) T1.413 Issue 2 and the European Technical Standards Institute (ETSI) Technical Report 328.

G.992.2, Splitterless ADSL Transceivers

This is a customer-installed, lower data rate version of G.992.1. It allows a defined service, with a specified data rate and supports a "best efforts" service (where the users' data rate is within a pre-determined range). The term "splitterless" means that there is no equipment at the customer location to "split" (separate) high frequency ADSL signals from the low frequency telephone signals. Bit rates up to 1.5 Mbps in the downstream (toward subscriber/user) direction and 512 Kbps upstream are defined.

G.994.1, Handshake Procedures for DSL Transceivers

Modeled after Recommendations V8/V.8bis (used by V.34 and V.90 modems), this is a start-up procedure to allow the support of G.992.1 and G.992.2 transceivers in a predictable and user friendly manner. Without G.994.1, unknown incompatibilities between different ADSL implementations could occur.

G.995.1, Overview of DSL Recommendations

This document provides the overview of the xDSL Recommendations and includes a system reference model.

G.996.1, Test Procedures for DSL Transceivers

This Recommendation is modeled after the V.56 modem test Recommendations. This standard offers procedures to be used by xDSL manufacturers and service providers to insure proper and compatible operation of the various xDSL implementations.

G.997.1, Physical Layer Management for DSL Transceivers

G.997.1 defines the physical layer operations, administrations and maintenance functions. G.997.1 is based on an Internet management approach (Simple Network Management Protocol, SNMP) and allows communications services providers (ILEC or CLEC) to manage and provision xDSL systems.

Developing xDSL Standards

At the ITU Study Group 15 Question 4 meeting on Digital Subscriber Line (DSL) standards, held in Melbourne Australia March 29 to April 2, 1999, more than 80 people from more than 50 companies met to develop worldwide xDSL standards. Work began on the new Very high speed DSL (VDSL) and Single-pair High speed DSL technologies (SHDSL) and the final drafts of the six ADSL Recommendations were reviewed.

Currently, maximum data rates for VDSL have not been agreed upon, but could be as high as 52 Mbps asymmetric or 26 Mbps symmetric over short copper loops. Several technologies have been proposed for VDSL, including single carrier and multi-carrier modulation.

SHDSL will provide capabilities including symmetric date rates, operation using repeaters to support more distant users, and remote powering to support backup operation. SHDSL could support peer-to-peer communications up to 2.304 Mbps in both directions. Determination (technical approval) for SHDSL is now planned for April, 2000.

xDSL Hurdles

Distance from a serving CO (Central Office) is a limiting factor. xDSL can reach from about two to five miles and is usually specified at up to 18,000 feet. These limitations cover about 80 percent of United States subscribers. The other 20 percent require repeaters on their local loops because of their distance from the CO.

Specific data rates may be lower than advertised because they are related to many factors: distance between you and your service provider (called "reach"), acceptable error rate, noise from customer premises wiring, and the electrical characteristics of the CPE (Customer Premises Equipment).

xDSL technologies are private line services, not switched. ISDN and xDSL compete in some applications, but ISDN's circuit switching capability means that users can connect to the public switched telephone network (PSTN) using ordinary dial telephone numbers as the address. Non-switched applications like xDSL use data addresses, predominantly IP addresses.

Line setup might be another problem. You need a dry (no dial tone), unconditioned copper pair of wires. You have to set up xDSL hardware before you can send or receive data. You must get the settings right before you can use it. Ready-to-go xDSL providers can (and do) send you a modem and instructions, or fix plugged-in CPE with remote management software.

Finally, xDSL providers are dependent upon being able to get copper pairs to you from the telephone company's CO. If there aren't any spare wires feeding your location, there'll be an additional cost for the telco to install new cable, and the work might take some time to get done.

xDSL: Present and Future

Dataquest, a market research firm, forecasts that 5.8 million xDSL lines will be installed by the year 2000. Compaq, Intel, and Microsoft are working with manufacturers to accelerate deployment of DSL Lite.

xDSL is being deployed in major metropolitan areas around the country. Look to alternate providers for full-featured xDSL services at very competitive prices. CLECs offer xDSL in combination with competitive local phone service over telco-owned copper pairs.

Computer Telephony Integration

Computer Telephony Integration and Applications

Computer Telephony (CT) and CTI (Computer Telephony Integration) are terms used interchangeably to refer to a group of technologies that coordinate the interaction of telephone and computer systems.

Computer telephony makes phone calls smarter by linking the computer mainframe containing a company's data and software programs, and the PBX (Private Branch Exchange) that controls the telephones.

PC-based communications servers can replace phone systems and transport voice and data over LANs (Local Area Networks). Major phone switch vendors are developing server-based phone systems, so that their phone systems can be attached to a PC network and work as a server to transmit and route data and voice.

Some communications servers are also IP telephony gateways, sending voice calls digitally using the Internet Protocol.

For more about Voice over the Internet Protocol, see the *Voice Over The Internet Or Intranets* chapter.

CT products vary in complexity depending upon whether or not they share telephony resources. A system that has sole control of a voice card and telephone trunk (any standalone CT application connected to a dedicated line) is much simpler in design and construction than a system that shares the control of trunks and processing with other systems and/or a human user.

Network interface technology lets computer telephony systems communicate electrically with outside telephone networks, the PSTN or a private network. Network interfaces interpret signaling coming across the trunks and circuits, provide data buffering, and also include surge protection circuitry.

Computer telephony systems interact with the telephone network in two ways. They can control how calls are established, reconfigured and torn down, the "call control" function. Call-control means making calls go to the right place, using CT as an intermediary or rule-based actor. A call control application enables a CT system to understand incoming call events (signals) and let it answer, make, transfer, conference, and otherwise manipulate calls.

CT systems can also send and receive information like facsimile, voice, tones, or data, called the media processing function. Media processing tasks often performed by CT systems include filtering, analyzing, recording, digitizing, compressing, storing, expanding, and replaying signals.

A computer telephony application usually performs some combination of call control and media processing functions.

Electronic call control and media processing functions have counterparts in ordinary telephone usage. Picking up a telephone handset, pressing touch tone buttons to dial digits, and then listening for tones that signal the successful completion of the call represent human call control functions. Once the call is established, speaking and listening are human media process-ing functions.

The first CT applications primarily performed media processing functions. Call control functions were limited to detecting a ring, answering the call, and hanging up after the message has been taken. The simplest voicemail systems answered incoming calls, presented a greeting, and then recorded the caller's message.

Newer voicemail and automated attendant systems can perform call control functions including call transferring, outdialing and paging. As the cost of telephone interfaces and computer tele-phony signal processing technologies have decreased further, computer telephony applications today can perform advanced media processing functions like voice synthesis, voice recogni-tion, and faxing.

Signaling

The most difficult accomplishment of a CT application is the proper signaling so that accurate and reliable call control is achieved. Recent advances in computer telephony have been improvements both in the underlying signaling connections and in the programming interfaces (APIs) which direct the application software programs that do the actual signaling.

Tone signals are processed by CT applications. Tone processing is a CT application's ability to receive, recognize, and generate spe-cific telephone and network tones so that a CT application can place a call and monitor its progress.

Most tone signaling used across networks today is in-band, from end-to-end. The best-known scheme in-band for terminal-equip-ment-to-network signaling is Dual Tone Multi-Frequency (DTMF), where the terminal equipment generates pairs of tones to represent each dialed digit.

Signaling between the worldwide telephone network and terminal equipment has not been standardized worldwide, so it's difficult to design CT terminal equipment that accurately interprets all of the various in-band signals that are sent across the worldwide telephone network.

Out-of-band signaling creates a direct digital information link between the telephone switch and the computer-based CT application. This approach is much more accurate than in-band signaling, but often requires proprietary equipment at either end of the communications link.

Out-of-band signaling is used via the D-channel associated with basic and primary rate ISDN service, and is the proprietary digital signaling method used between PBXs and digital telephone sets. The switch-to-switch signaling protocol called Signaling System 7 (SS7) also uses out-of-band signaling. Many of the CT links available with PBXs use out-of-band signaling. Out-of-band signaling is also used between switches across the PSTN.

Application Programming Interfaces and Call Control

The CT connection between the telephone system and the database it communicates with is called an application program, application, or app. Apps are software programs, running on the computer telephony-based PC system, that translate back and forth from between the phone system and the CT resources.

An application programming interface (API) is the design mechanism through which the CT application software manipulates the telephone resources. APIs let non-technical types program and maintain CT systems by providing easy-to-use software interfaces to CT apps.

First Party and Third Party Call Control

The relationship between a CT application and the amount of control it has over a telephone line is classified as either first-party or third-party call control.

First party call control gives PC users hardware control of their telephone set from their computers using a direct physical link between the desktop PC and the telephone set. With first party API software acting as the interpreter, telephone control functions are handled by unified messaging software running on the PC, while the CT software running on the PC provides the call routing functions.

Instead of controlling a telephony device inside or a phone connected to, a personal computer, a third party call control application controls a shared telephony resource like a key system or PBX. Third party call control lets a CT system talks to other systems, monitoring and controlling calls not only to and from the desktop, but also between desktops.

Proprietary APIs for first-party call control were first developed by modem, voice board, and fax board manufacturers to support their own products. The only API to achieve de facto standards status was the Hayes modem command set, which included basic functions for dialing and hanging up telephone calls.

In the 1980s many of PBX vendors opened up their switches to programmable CT links. This effort was called the Open Architecture Interface. Under OAI, CT applications had to be programmed for each different vendor's phone system; there was no standard application interface.

The first third-party APIs were developed in the late 1980s by computer manufacturers to support applications running on their own systems; IBM's CallPath API and Digital Equipment's

Computer-Integrated Telephony (CIT) are probably the best-known examples

CT took a major step forward in the 1990s with the introduction of two call control APIs that weren't linked to any individual equipment manufacturer, the Telephony Services API (TSAPI) developed by AT&T (now Lucent) and Novell, and The Telephony API (TAPI) developed by Microsoft.

TSAPI (Telephony Services API)

The TSAPI third-party call control API was specified by Novell and Lucent, supported by a large number of phone system manufacturer's PBXs and KSUs into the mid 1990s, and was the first widely implemented third party call control API. It is supported by most vendors now, but is not being used in new CT products.

TAPI (Telephony Application Programming Interface)

TAPI, introduced in 1993 by Microsoft and Intel, is a telephone application program interface for first- and third-party call control.

Microsoft TAPI is a client/server protocol that ships with Windows and allows computers running Windows 95, 98 and NT to control a shared telephony server. It used best when desktop PCs are communicating with Windows NT-based telephony servers or communications servers. It doesn't allow control of non-Windows telephony servers, which leaves out most telephone systems.

At its "basic services level" TAPI manages all of the signaling between a computer and phone network, including dialing, answering, and hanging up a call. At the (optional) supplementary services level, TAPI can also hold a call, transfer it, conference, and so on. Supplementary call control features are the usual province of PBXs. It is left up to the PBX vendor or comms switch device vendor to decide whether or not they want to support supplementary TAPI, and then to hand over those controls to the CT system.

Windows 95 looks at voice as being just more one media data type. Control over telephony media, like voice and speech recognition, is accomplished by Microsoft's media APIs, not TAPI. Playing and recording voice files use the WAVE (multimedia wave audio) API, for instance. This means that TAPI does not support multimedia communications, so voice, data and video must be managed externally to the PC by another protocol. Media processing requires multiple machines, running different data management programs.

CT-Connect

CT-Connect is Dialogic's proprietary third-party call control API. CT-Connect supports TAPI, TSAPI and other proprietary CT link protocols. Requests from all of the APIs is translated into into Dialogic's common function set.

Dialogic's CT-Connect, like TAPI, is based on the European Computer Manufacturers Association's (ECMA) Computer-Supported Telecommunications Applications (CSTA) specifications.

Sun Microsystems' JTAPI

JTAPI is an extension to Sun Microsystem's Java programming language for third party computer telephony applications. JTAPI's advantage is that it provides a cross-platform interface; JTAPI supports multiple operating systems, where TSAPI and TAPI limit developers to 32 bit Windows operating systems like Windows 95/98.

COMPUTER TELEPHONY APPLICATIONS

Computer telephony systems fall into a few functional categories. The most popular CT technologies have their own chapters in this book: *Communications Servers, Voicemail, Interactive Voice Response (IVR) and Auto Attendants, Voice Over the Internet Protocol (VoIP), Unified Messaging and Enhanced Services.*

Automatic Speech Recognition (ASR)

Automatic speech recognition (ASR) technology (sometimes known as voice recognition) reliably recognizes human speech, such as discrete numbers and short commands, or continuous strings of numbers, like credit card numbers.

Speaker-independent ASR can recognize a group of words (usually numbers and short commands) from any caller.

Speaker-dependent ASR can identify a large vocabulary of commands from a specific speaker. Speaker-dependent ASR is used primarily for security systems and recently, for hands-free work environments.

As microprocessors get smaller and storage media gets larger and cheaper, larger libraries of speech are possible every day. Speech recognition holds great promise as a replacement for touch tone-based IVR systems.

Text-to-Speech (TTS)

Text-to-speech (TTS) technology generates synthetic speech from text stored in computer files, providing a spoken interface to frequently updated information. Callers can access computer-based files, as often as they like, through their telephones.

TTS gives companies that would normally mail out lots of paper-based information an electronic way to disseminate the same quantity of information very inexpensively.

Switching and Conferencing

Switching and conferencing technologies handles the routing, transfer, and connection of more than two callers. These capabilities, once the province of private branch exchanges (PBXs) and proprietary switches, are now available on expansion boards for PC-based CT systems.

Multi-Media Processing

Computer-based media processing hardware is pretty simple as long as each telephone line has a dedicated set of hardware resources. Media processing hardware gets considerably more complex, however, when applications have to reconfigure resources on-the-fly.

Reconfigurable on-the-fly means that any incoming call on any channel must be able to be switched to digitizers, playback units, recognizers, fax processors and analog interfaces in any combination. All of these processes won't, and shouldn't fit onto a single circuit board, for expansion reasons, and so as not to have a single point of failure in the system.

New CT APIs and resource models are being developed to allow shared media processing resources on shared servers.

Cross-vendor efforts dedicated to address shared media processing include the Multi-Vendor Interface Program (MVIP) and the Enterprise Computer Telephony Forum (ECTF)'s SCSA working group, each of which has developed proposals relating to software architectures and APIs for shared media processing resources.

MVIP (from the GO-MVIP organization), and Scbus (developed by the SCSA working group), specify time-division (TDM) based buses for connecting the talk paths and a separate communication mechanism for coordinating the subsystems.

The MVIP/SCSA vendors are now offering fully-distributed PC-based telephony systems capable of a thousand+ ports, including media services, that work without a PBX. Mitel, one of the world's largest PBX vendors, has already endorsed this distributed, non-PBX CT system.

SCTP (Simple Computer Telephony Protocol)

SCTP is a comparatively simple, vendor neutral signaling protocol designed to run over an IP network. It is in the process of being submitted to the IETF for adoption as a public domain standard.

SCTP is consistent with other Internet protocols, and also employs syntax similar to that used in CGI (common gateway interface) transactions.

SCTP provides telephone companies with an Internet equivalent of an ISDN D channel and adds features including third party call control. The signaling channel (SCTP session running over the Internet) is associated with a separate voice path. The Internet connection works as an out-of-band signaling channel alongside a connection to the PSTN.

Most computer telephony software has been limited to managing calls using an office PBX. Computer telephony APIs (namely TAPI and TSAPI) have historically been difficult to support since most telecom equipment is not PC-based. A dedicated computer telephony server was often required to translate between phone systems and Windows-based computers. Moreover, the most popular APIs on the market are Windows-platform specific.

SCTP enables client/server call control and station management applications on devices ranging from Java-enabled smart phones to network computers to traditional desktop workstations (running the OS of your choice).

It will be easier for telecom equipment vendors to add network CT capability to their equipment, and vendors will be able to program their equipment to talk directly to clients using the SCTP interface. SCTP can easily be supported on telecom equipment from key systems to national telephone networks.

Present and Future Computer Telephony

The traditional (and relatively slow) way to connect computers to phone systems uses analog ports on the phone system to connect to the CT application. One adjunct PC may be needed for each CT application. A typical installation composed of voice mail, ACD, and a fax-on-demand application would require three separate PCs.

Digital CT applications can run on the same PC that contains all of the interface ports and do not use analog ports to connect media processing channels to the switch (requiring analog interface cards, and a ring generator or DTMF receiver).

Not only is call processing using digital phone ports much faster (faster ringing, on- and off-hook transitions, and dialing), but the voice quality is also better since the signal remains digital all the way from the phone system to the PC's processing board. There's no more "pop and click" when a port hookflashes to transfer a caller.

Computer telephone applications are not restricted to the traditional forms of telephone systems based on switches, transmission circuits, and telephone instruments. See the <I> Voice Over The Internet Or Intranets</I> chapter for more about the latest CT application: Ethernet-based H.323.

Voice Over the Internet or Intranets

The Internet Protocol

Internet Telephony, VoIP (Voice over Internet Protocol), IPT (Internet Protocol Telephony) and IP (Internet Protocol) Telephony are a set of CT (Computer Telephony) technologies and products whose markets are new in the last couple of years. Their commonality is the IP protocol, which is a digital, packet-switched set of instructions for sending voice and fax information over the Internet and over private intranets.

The terms IP telephony and Voice Over IP (VoIP) do not generally refer to the same technology as the term Internet Telephony. IP telephony, IPT and VoIP generally mean voice calls transmitted over a packet-based data network (either public or private), using the Internet Protocol. Internet Telephony implies that the call actually goes over the public Internet at some point.

IP telephony technology breaks analog phone calls into digital signals and sends voice calls over the public Internet or over a private intranet using the coding mechanism (protocol) called IP. An IP address identifies a particular computer on an IP network, like a telephone number identifies a particular telephone. You have an

IP address when you dial into the Internet or get to the Internet through a corporate network.

The traditional phone network (PSTN) is circuit-switched. Each phone call creates a circuit between one phone and another. A circuit-switched call travels a fixed path. IP telephony is a packet-switched technology. During a packet-switched call, all of the packets associated with the call try to reach the same destination, but they don't all have to follow the same path.

Packet-switched technologies, like IP telephony, convert voice calls from analog to digital and then divide the calls into chunks called packets.

Each packet is wrapped with addressing information indicating who sent it and where it is going and a sequence number so that it can be reassembled at the other end. A dedicated (or circuit-switched) link between the callers isn't established.

Routers and servers at each end of an IP telephony call read the addressing information and direct the packets to their destinations.

Internet Telephony

The Internet does not guarantee that a specific amount of data will arrive within a specified time frame. Packets are routed first-come, first-serve. If enough data is pushed through the same path at the same time, there may not be enough bandwidth for all of it.

On the open Internet, packets can be lost or delayed, particularly during periods of congestion, because packets are discarded when errors occur during transmission. Lost, delayed and damaged packets cause deterioration of sound quality.

Packets on a packet-switched call can use different routes over the network to get to their destination, so you can have two problems when the network gets busy. One is latency, where some packets arrive a little later than other packets. The worst scenario is that some packets arrive very late at their destination or not at all. Lost packets, (called packet loss as a measure of data communications integrity) make voice calls over the public Internet unpredictable, and often unsatisfactory in comparison to a circuit-switched voice conversation.

PC to PC voice over the Internet products are generally not used at a corporate level. Private IP networks, however, are installed and in use because they are not subject to the delays inherent in the public Internet. On a private IP network, there is resource allocation authority in the form of a network engineer or manager, so if you need more bandwidth for heavy traffic times you can allocate it dynamically.

Why IP?

IP telephony is a cheap alternative to traditional long-distance phone calls. The long distance communications market is extremely price elastic, so as prices fall, traffic increases.

Right now, IP telephony is inexpensive because the price incentives for the development of the technology depend on its ability to bypass regulatory regimes and tariff structures. The Federal Communications Commission currently doesn't consider voice conversations over IP networks to be telephone calls, so you don't incur long-distance phone or fax charges if you contact people directly over an IP network.

IP networks are treated by many foreign governments as value-added networks and therefore are not subject to accounting rate settlements. IP voice service providers can charge less than traditional long distance carriers who are subject to tariffs.

IP telephony switching makes the carriers' existing imbedded networks more efficient. The way the call information is sent has no bearing on the type of call being transmitted, so voice, data and video calls are treated the same by the network. The equipment and wiring involved can be similar.

International callback providers were the first to develop IP telephony-based services to protect their markets. The theory was that their customers, so price sensitive that they are willing to dial additional digits or suffer other inconveniences to save money, would readily switch to a cheaper technology, even though it is somewhat difficult to use.

Big long distance providers also implemented IP because their profits were threatened and they understood the opportunity to make money by using IP themselves.

Types of IP Telephony

There are basically five functional types of IP telephony communications: PC to PC, PC to phone, phone to phone, premises to premises and premises to network.

PC to PC Internet telephony is the oldest IP telephony technology. To use IP telephony without telephones, you need a computer with a microphone, a headset, a sound card, desktop software, and access to an IP network.

Each end's PC has to be on and ready to accept a call, and delays of more than half a second between voice transmissions over the public Internet are typical.

Individual-PC Internet telephony software is accessible and inexpensive; Microsoft NetMeeting, available free, is a popular product for Internet telephony.

PC to Phone The call originates from a PC, goes through a dedicated or dial up Internet connection, and terminates at a normal phone at the far end.

The ISP (Internet Service Provider) that the PC dials into provides a gateway that switches the call to the Public Switched Telephone Network (PSTN), and then the call is sent to the remote phone. This type of call is also subject to Internet-based delay.

Phone to Phone This requires multistage dialing. You dial an access number into an IP gateway, then you enter a PIN number and the number you want to reach. The call is sent to a gateway close to the recipient's location, which hands it off to the local PSTN for completion.

The last two ways to make an IP telephony call require IP Gateways. Gateways (or a third-party service that provides the gateway conversion function) are used to route regular phone calls over a private (or to the public) network, using the IP protocol.

Business Premises-to-premises versions of an IP telephony call use IP gateways between the company PBXs with a private or public IP network in the middle.

Premises-to-network calls originate onsite and are sent to private or to public networks via an IP gateway.

IP Telephony Gateways

Gateways connect two dissimilar networks and perform any network or signaling translation required for interoperability. They also translate protocols for call setup and release, convert disparate media (audio, video and data) formats, and transfer information between H.323 and non-H.323 networks.

An IP telephony gateway makes it possible for you to receive cir-cuit-switched calls, switch them within your company as packet-based data, and then convert them back to voice to return them to the outside world through your PBX.

Internet telephony gateways take voice (or fax) calls from the cir-cuit-switched PSTN and place it on the packet-switched Internet and vice versa. VoIP gateways are local to both parties so that both parties are charged only for local calls.

Gateways are PC-based communications servers with high-end voice boards inside. The voice boards digitize voice conversations so that your data networks can transmit them as packets. Your data network sends these packets to another gateway, which converts them back into audio signals. You hear these signals as voice con-versations at the other end.

At present, most IP telephony applications are run over private networks where the bandwidth, and therefore the quality of the transmission, can be controlled.

One of the best uses for IP telephony will be high-speed digital delivery of local dialtone. A high-speed circuit from an Internet service provider and a H.323 gateway in your office makes the data network look like a bank of analog telephone lines. You con-nect your office telephone system to the gateway and it thinks it is connected to Central Office trunks.

The ISP (Internet Service Provider) gateway in the local switch-ing center must be the same as yours so that calls can be directed back out onto the public telephone network. You can order one high capacity circuit to handle all of your voice, video and data networking services.

A gateway converts calls coming in from the telephone network into packets suitable for transmission over the Internet, and vice versa. The key feature of these arrangements is that the gateway must be connected to the telephone network in a way that allows it to receive traditional telephone calls for conversion to packets in the Internet Protocol. These connections will trigger certain regulatory consequences.

The FCC has indicated that some technical arrangements that use the Internet to transport voice calls will be classified as regulated telecommunications services, and therefore subject to tariffs. There are basically three situations in which this might occur. In each case, Internet telephony is provided by an IP gateway

One option is to make the gateway available to callers as a local call. A typical call to an ISP is a local call. When the gateway answers with a computer tone, the caller enters an identification code for billing and then dials the destination number.

The call is carried over the Internet to the destination and there the translation process is reversed so that the destination telephone rings.

The gateway's telephone lines (with the number you call to reach it) would not be classified as end-user business lines, which is how an ISP's normal dial-in lines are classified. Instead, those lines would probably be classified for regulatory purposes as a form of "access" (the type of service that long distance carriers pay local exchange carriers for). Access charges would mean much higher monthly charges for the ISP, and presumably gateway customers.

A second option is to make the gateway available to callers through a toll-free "800" or "888" number. You dial the number and after the gateway answers, enter a code number, then dial the destination number.

In this configuration, the gateway provider will be charged a per-minute rate by the long distance company that supplies the "800" or "888" service. The LD provider, in turn, pays the local exchange carrier a per-minute access fee for the service of routing the dialed 800 or 888 number to the correct long distance carrier.

Finally, the gateway itself could be set up as a normal long distance carrier. Using this method, you would dial a "1010XXX" code before dialing the destination telephone number. The local telephone network would capture the information needed to bill for the call, and either pass that information on to the gateway or bill you on behalf of the gateway. In either case, the local telephone company would charge the gateway a per-minute "access charge" each time someone connects to it through the local exchange network.

The fact that gateway-based IP telephony arrangements might be regulated does not mean that they cannot offer good service at rates far below traditional telephone rates, particularly in the case of some international calls subject to high "accounting" and "settlement" rates imposed by non-US nations. It does mean, however, that-while the regulatory rules applicable to gateway-based IP telephony are not yet settled-firms seeking to provide such services cannot conclude from the mere fact that the Internet is involved that they are automatically shielded from the normal regulatory consequences of providing telephone service.

Organizations are starting to allow web surfers with Internet telephony software to connect directly to their employees by using an Internet telephony gateway. The Internet telephony gateway feeds calls from the Internet directly into their ACD (Automatic Call Distribution) system.

A CLEC (Competitive Local Exchange Carrier) that provides gateway service installs a large enterprise-class or carrier-class gateway with good scalability. If their service area is large enough, they install a gateway at each major point of service, then route calls between them and hop off to the PSTN at the closest (or lowest cost) point to complete the call.

Private companies that want to use VoIP might put an enterprise-class gateway at each major site, and much smaller gateways in branch offices.

Most gateway manufacturers offer a variety of options for each side of the hookup-two to 16 analog voice ports or up to 24 digital voice ports on the customer side, and a 10 or 100 Mbps Ethernet connection on the other.

IP Telephony and SS7

The PSTN has fancy features called Intelligent Network (IN) services that are made possible via the Signaling System 7 (SS7) protocol in the Central Office switch software. SS7 is used for basic call setup, management, and tear down, and it allows database query for local number portability, mobile (cellular) subscriber authentication and roaming, virtual private networking, and toll-free service.

To interoperate with the PSTN, an IP telephony switch must support the Signaling System 7 (SS7) protocol. If an IP telephony switch lacks SS7 capabilities, PSTN databases which store routing numbers associated with toll-free and ported local numbers can't be accessed, and the IP calls can't be switched.

Gateway vendors are working at developing SS7 integration for their IP telephony solutions to ensure PSTN interworking, a critical first step in the move to a reliable public IP telephony network.

VoIP Service Providers

Usually, IP voice service providers sell international service to and from the United States from gateways co-located with a PSTN switch. US-originated calls get routed by the IP voice service provider to their IP gateways, generally using a toll-free number.

In some cases, there is sufficient traffic in the vicinity of the provider's switch to cost justify using local CO trunks or ISDN PRI lines to connect users directly to their gateway.

IP Telephony Applications

Fax Over IP
Fax over IP lets you send your fax as a packetized data transmission instead of as a circuit-switched voice phone call.

IP faxing means sending faxes all the way, or part of the way to their destinations over a IP-based LAN, private WAN, or the Internet. IP-based faxing is attractive to businesses for free intracompany and cheap international faxing.

There are two ways to do this: store-and-forward and real-time.

Store-and-forward faxing converts the output signal of a fax-appliance (any fax machine or PC-based fax card either in a desktop system or a fax server) into a data file, transmits this data file across the Internet then prints out the file at the other end, either direct to a network printer or by calling up a local fax machine. A store-and-forward fax-over-IP box comes with a small hard drive for storing faxes, an Ethernet port to plug into your WAN and four loop start ports to plug in standard fax machines or lines from your PBX.

A variation on the store-and-forward theme involves sending faxes to a central Web server, and letting the recipients review them using their Web browsers.

Real-time faxing over IP resembles traditional fax. Here, specialized buffer boxes establish a real-time connection between two fax appliances across the Internet, and fool the fax machines into thinking they're talking over a regular phone connection.

Fax over IP service bureaus let you send faxes from any fax appliance by making a local or 800-number call. Fax over IP gadgets and servers hook up to your LAN and phone extensions; just call them from any fax machine, press some keys to identify your destination, hit SEND, and you're IP faxing.

One application for faxing over Internet has appeared in the international fax marketplace. An international fax over IP service with POPs can save big bucks, often a 40% or 50% percent savings compared to an international call from your fax machine. Fax over IP services to foreign countries can also ease the problems of international fax machine interoperability.

Faxing over an IP intranet means you don't have to worry about Internet delays and bandwidth bottlenecks, because you control (and pay for) the connections and bandwidth.

This technology is still new, so make sure you have the latest version of all related board drivers, software, etc. You can look on the Web for software and driver update news.

Another fax over IP application involves a regular FOD (Fax on Demand) system that integrates with Web sites. Visit a company's Web site, click on the button, enter your fax number, and receive a FOD document at your fax machine. Since the products use regular FOD technology you can use a single repository for all your documents. Better yet, Web surfers don't have to wait for their slow modem connections to deliver the document.

Conferencing

There are IP telephony software applications (generally H.323 compliant) that enable people to talk and share data simultaneously over a single IP connection, conduct videoconferences using IP and share documents over that same connection. Visual interaction and the ability to work with a document despite physical distance is of tremendous value to companies that would otherwise conduct a voice-only conference.

Conferencing is a great application for IP telephony. In addition to obvious cost savings compared to conferencing using the PSTN, IP telephony-based conferencing gives control over the communication to the individual company. Traditionally, large teleconferences and most videoconferences are handled by outside companies for most organizations.

Video calls have bandwidth requirements from 28.8 kbps to over 300 kbps depending on the quality of the image being transmitted. Broadcasting high quality video sessions requires at least 1.5 Mbps per session.

Video calls where the primary image being displayed is the face of the person placing the call usually use the least bandwidth. Most of the information displayed on the screen remains the same from frame to frame, so the bandwidth required is less than for full-motion video.

Unified Messaging

A key application for IP-based intranets is unified messaging implemented so a user can pull down email, pager, voicemail and fax messages from a corporate web page, checking messages just once and managing them together.

For more about Unified Messaging, see that chapter in this book.

VoIP in Call Centers

IP telephony offers real benefits for call centers. IP telephony is an option for reaching live agents in addition to text chat or Web callback.

The most common IP telephony applications for call centers let callers talk to an agent over the same connection they've used to dial into the Web. This means the call center agent can monitor the caller activity as the website is explored. Monitoring is especially powerful in help desk and sales applications. If you have a clientele that is comfortable using a computer to call your business, there are many call-me button IP telephony software products you can use.

H.323 and VoIP

H.323 and some associated protocols create a standard procedure for setting up a phone call (voice or video) between devices, transmitting data between them in real-time, and then tearing the call down when it is done.

H.323 also defines how to determine the best way to place a call by choosing the best compression algorithm to use for each call. It also provides support for Internet to PSTN gateways that route calls to any traditional phone number.

H.323 is designed to be a technology for multipoint-multimedia communications over packet-based networks, including IP-based networks. It can be applied in a variety of ways-audio only (IP telephony, or VoIP), audio and video (videotelephony), audio and data, and audio, video and data. Right now, IP telephony is the most-implemented application.

In the near future, the multimedia aspects of H.323 will be important in delivering combined voice, video and data communications services over IP-based networks.

The most important feature of H.323 is that "Internet phone lines" can be used with any type of Internet connection provided there is enough bandwidth to carry the call.

The H.323 protocol tells the gateways on either end of the IP telephony call how to handle each end of the conversation. If an internal or external IP telephony call comes in from a PC, the protocol stack lets the gateways know which H.323-compliant software is being used and what to do with it.

You can add switched hubs to the IP network so that local area network traffic hitting the file servers (the computers that translate and switch the call data) does not affect the quality of your voice calls. (Voice does not use much bandwidth; about 20 kbps using common compression techniques, down to as low as 8 kbps with other compression techniques.)

As more vendors support the H.323 standard, the proprietary restrictions of person-to-person and gateway-to-gateway IP communications will disappear. Version 2 of the H.323 standard defining packet-based multimedia communications systems, formalized in January 1998, defines additional requirements for VoIP. New features currently being added to H.323, like gateway interoperability, will result in a new version of the standard in 1999 or 2000.

The Future of IP Telephony

There are several IP telephony interoperability issues that need to be resolved. Network vendors are not yet supplying interoperable gateways. Existing products often use proprietary protocols and do not work well with products from other vendors. Service providers aren't comfortable deploying proprietary solutions.

IP telephony standards, such as H.323, are evolving rapidly, but important functions, such as inter-domain communication, network management, and billing, have yet to be defined. Most IP telephony products on the market today do not support full Signaling System 7 (SS7) capabilities and therefore cannot interoperate with the PSTN.

As new IP signaling protocols are standardized and implemented in network routers, the quality of voice over the Internet will become equivalent to that of the PSTN. Until then, VoIP technology will be used primarily by large organizations and public carriers over dedicated networks (rather than over the public Internet) where quality-of-service can be monitored and controlled.

The next wave of potential users of IP telephony that are attracted to lower prices will adopt the technology only after it is as convenient as traditional voice services. Although the gateway market has evolved at a tremendous pace, the IP Telephony gateways available as of late 1998 are still lacking.

In 1999, several service providers will introduce IP telephony and fax services that no longer require the end user to dial any differently than they would for a regular phone call (called single stage dialing). This and other features of the traditional voice network will be available for IP telephony networks as vendors introduce SS7 integration.

If IP telephony is going to be generally adopted, it will have to be price competitive with existing circuit-switched voice technologies. The cheapest large-scale private IP telephony networks still cost roughly four times as much as an equivalent circuit-switched network.

IP telephony gateways sold for an average price of over $1,800 per port in 1997 and will be down to about $600/port in 1999. The same port in a traditional carrier network costs less than $100. This is why traditional carriers continue to have objections to large-scale deployments of IP telephony in lieu of circuit switching.

At the enterprise, or individual company level, IP telephony networks have not taken off because appropriate pre-packaged solutions aren't available. The technology is still immature and

IP telephony installations are one-by-one and customized, and potential users are skeptical about the quality of service. Existing telephone networks work all right, and companies need a good reason, usually big cost savings, to change technologies.

Unified Messaging

Unified messaging means that your PC acts as a universal inbox for all of your messages: voice, fax, email and attachments. Click on an email message, and it is displayed in your email editor. Click on a fax message, and it is displayed in a fax-viewing program to display and print. Click on a voice mail message and it is played back over your PC's speakers.

Unified messaging lets you prioritize your communications and take control of your work.

The type of message (voice, fax, email, whatever) is identified, as is the length of the message, the time it was received and the sender. You can listen to and/or view messages in any order and sort messages by type or age.

Unified messaging applications often work with caller ID information. The Central Office (CO) passes the phone number of the caller through a hardware/software CT application installed in the PC to a PIM (Personal Information Manager) software program running on the PC. As the call is ringing, the computer accesses the PIM database and pops the caller's file onto the computer screen.

Standard analog business lines transmit Caller ID using inband signaling between the first and second full ring cycles.

For regulatory reasons, most phone companies do not offer Caller ID services on T-1. ISDN usually does provide caller ID signaling, so most CT systems use ISDN when caller ID information is part of the product.

There are two ways universal inboxes are built. Some use Microsoft Exchange, a message server that is designed to handle many media types. The other uses an Internet standards-compliant mail server like NT Mail or Netscape Mail.

Microsoft Exchange is widely supported as part of the Windows 95 and Windows NT operating systems. Many voice mail, fax server and email software vendors already support Exchange and MAPI (Microsoft's messaging interface). An Internet-compatible product that integrates well with standard Internet mail servers means that end users can use browser software on their desktops.

Internet standards-compliant software for unified messaging is not as popular as Microsoft Exchange-based systems, but they are open interfaces. Varied and different computers and systems interface with IP (Internet Protocol) based systems. Not so with Microsoft products.

If you can, choose a unified messaging system that supports both Windows and IP-based messaging.

Features and Benefits of Unified Messaging

Instant callback
If the sender of a voicemail message can be identified and the CT system can dial out, calls can be returned with a single click of your mouse.

Call control

Unified messaging systems let you set up screening and switching paths on an ad-hoc basis, specific to caller identity, time and number called.

Call sorting

You can set the system to send certain calls to voicemail and others to your screen. Once the call pops, you may decide to answer it, or you could decide to drag and drop it to voicemail after all. Or, monitor the call and break in if you decide the call should be answered based on the message the caller is leaving.

Call association

Voice, fax, and email messages can be associated with each other. Send a fax message attached to a voicemail message. Attach audio files to email messages.

Superior voicemail

Unified messaging makes voicemail documentable. Voicemail messages are usually deleted by their recipients once they are listened to. PC-based voice recordings can be archived and reviewed in an orderly way.

Unified messaging technology makes voicemail really work. By adding on-screen reviewability and archivability, the ability to file messages with related items on the computer, multimedia mail and other features, unified messaging turns "voicemail" into a truly workable deferred messaging medium.

A well-installed unified messaging system should make a lot of the "call me back" messages disappear. It should replace them with messages that contain real information that you can prioritize, file and review again later.

How Unified Messaging Works

Unified messaging systems are mostly offered by voicemail manufacturers so are usually built on a voicemail platform. Unified messaging is presented as an optional feature or modular upgrade to a voicemail product.

Unified messaging systems look like voicemail systems; they usually run on a PC platform under Unix or OS/2. The PC contains voice cards, and is interfaced to your switch by traditional voicemail methods (analog or digital ports, and inband or out-of-band signaling).

A LAN network interface card (NIC) goes in the voicemail box with drivers, and is attached to your network. The system gets configured to talk to other servers (e-mail, MAPI, Microsoft Exchange, etc.) as possible, appropriate, and useful. Then the voicemail/unified messaging system gets synched up with email servers' address lists, usernames, passcodes, etc., so that the voicemail system can access user e-mail.

You'll also need to install client software on desktop PCs, and configure the system. You can enhance PCs by adding sound cards as needed, but most unified messaging systems will play voicemail back through the telephone.

Unified messaging is not end-user-installable. It's relatively complex; it has to interoperate with your switch and your network. There are lots of "fine tuning" things that have to be done before the system will work. It's really a VAR/integrator job.

The ultimate answer for unified messaging is for all mail to travel digitally over an ISDN or ATM or Internet network, carrying message surround information (time, date, originator, length) with it, in standard, computer-readable format. That technology is not quite here yet.

Enhanced Service Platforms

International Callback and Calling Cards

Enhanced service platforms are used to provide international callback (dialback), debit and credit card calling and prepaid long distance services. These hardware and software Computer Telephony (CT) products combine switching functions and database-querying functions.

International Callback

The United States has a competitive long distance market so international rates to most countries are the cheapest in the world.

Countries foreign to the US sometimes have state controlled telephone service that profit their governments. These countries have been reluctant to reduce their rates and therefore their profits.

Callback has provided a simple alternative to using high cost local services for over ten years, involving no change to equipment, no new installations and no new technology to learn.

International callback provides overseas customers with US dial tone, through which they can originate calls at much lower rates.

Here's how traditional international callback works:

1. You call a telephone number in the United States. You let this number ring once, and then hang up.

2. A computer in the United States calls you back at your offshore location. You enter a PIN number, and then the destination telephone number.

3. The computer in the US originates a call from the US to your destination, and then connects you through.

Technically, both calls (the call from the US to you, and from the US to your destination) originate in the United States, and are billed at much lower rates.

In recent years the profit margin of traditional callback has shrunk. Deregulation and privatization have spread around the world, leading to reductions in foreign-originated telephone rates and many foreign countries have declared callback unlawful within their borders. Most recently, callback profits have been squeezed even further by assessments levied under the Federal Communications Commission's (FCC's) Universal Service Fund (USF) according to the Telecommunications Act of 1996.

Although pressures on callback are likely to continue, the FCC is currently considering two petitions that have the potential to lighten these pressures.

Under the FCC's Universal Service program, callback providers, as well as other types of service providers, currently pay 3.9% of their interstate and international revenues to support the USF. The program requires that any provider offering interstate telecommunications services pay into the fund.

The USF assessment strikes callback companies particularly

hard. Most derive the their revenue from foreign originated calls and only a small portion from domestic, interstate traffic.

Nevertheless, because many callback providers have some domestic traffic, the FCC's rules require them to contribute to the USF based on their total international, interstate and intrastate revenues. This can result in a callback company having to make USF contributions which are several times greater than the domestic interstate revenues that triggered the requirement.

To make matters worse, overseas competitors of callback companies are often entirely exempt from the USF contribution, giving them a significant competitive advantage in the market.

One important recent innovation to come out of the callback industry is a device that tricks office phone systems into seeing an international callback network as if it were providing local dial tone. These devices, when they sense an outgoing call, signal the international callback network, wait for a return call, and then dial the destination telephone number into the callback switch. The whole process is completed in a matter of seconds, and the PBX to which the box is attached thinks it is placing a call over a normal direct dial line.

This greatly alleviates ease of use issues which have plagued callback users: having to hang up, wait for a return call, dialing a PIN then dialing the destination number. Callback service providers will likely provide the new equipment to business customers. The easier they make it for you to use their network, the more calls you are likely to place through it.

The biggest "legal" problem with callback involves "code calling," using the PTT (Postal, Telephone, and Telegraph systems owned by the non-US government) to make an (unanswered) long distance call when signaling the US-based provider to start the callback process. Callback platform vendors have worked out several ways to avoid this unfair and sometimes illegal use of PTT services. The newest systems place PC-based local switches in the cities they serve, connecting them to the Internet or to packet-data services. Clients desiring callback service make a local call to the switch, enter a PIN code, and hang up. The foreign switch signals the stateside provider across the data network, identifying the subscriber and starting the callback process.

Another, simpler "local node" strategy can be used where the need is consistent enough to amortize the cost. In this case, the US provider calls the overseas node and keeps the line open; when a subscriber calls the node locally, US-based dialtone is immediately available.

One of the biggest deal-breakers facing conventional callback is its inflexibility. In order to use the service, you have to be at your account's pre-programmed number. But new callback systems are managing to fix this problem in a variety of ways.

Callback products are evolving that offer useful services including voicemail, "find me" telephony, calling card services and speed-dial.

Some products employ a front-end IVR on the US-based platform. Subscribers can call their regular number, and the server will eventually answer. They can then use the IVR to reconfigure their accounts: entering new callback numbers (where they're staying tonight) and recording greetings that the system will read to human attendants to facilitate call completion.

The most sophisticated systems use the Web for new account setup and reconfiguration. The host service can be mounted on any Web site. Simple CGI scripts or Java applets communicate with the Stateside platform over TCP/IP.

There are hundreds of international callback service providers, some of them very specialized. There are a few large players who offer competitive rates worldwide and there are dozens of smaller companies that concentrate on specific geographic markets. If you make calls from a relatively large number of countries, you'll probably want to use one of the larger service providers; if you call a specific list of countries, you'll probably save money by going with one of the firms specializing in the countries you're dealing with most.

Calling Cards

Calling card applications require a switch to handle inbound calls and to hand off the caller to a carrier, and a host computer that processes card ID data and updates account records.

Pre-paid calling cards are debit cards. You buy them for a specific amount of money and get a specific number of minutes for the price. The per-minute rate of calling cards are usually higher than credit calling cards and calls made from your home or office. Debit cards are most economical for consumers who tend to make short calls away from home.

Consumers should consider the following before purchasing a pre-paid calling card:

What is the actual cost per minute? What is the card's geographic coverage? (Not all cards allow you to call internationally.) Finally, how reputable is the company? Many companies have recently gone out of business, leaving their customers with unreimbursable, unused minutes.

If you have many people using company calling cards, you can control their usage several ways. For one, you can limit the number of people with access to company calling cards. Second, you can issue a unique account code to each card (or a unique card to each individual) so that you know how much each individual is spending. Finally, you can place dollar limits on monthly usage.

Reference and Templates

Key Sheet

Name	Department	Location	Cable#	Phone Type	Primary Extension	Secondary Extension	COS (class of service)	Feature Table	Forward Target Busy	Forward Target RNA	Other Connections

Telecommunications Organizations

American National Standards Institute (ANSI)
American National Standards Institute
11 W. 42nd St, 13th floor
New York, NY 10036, USA
Tel: (212) 642-4900

ATM Forum
Focuses on speeding the development, standardization and development of ATM (Asynchronous Transfer Mode) products.
Mountain View, CA
(415) 949-6700
www.atmforum.com

Bell Canada
Building Network Design
Floor 2, 2 Fieldway Road
Etobicoke, Ontario
Canada M8Z 3L2
Tel: (416) 234-4223
Fax: (416) 236-3033

BICSI
A telecommunications cabling professional association. Offers education, and administers the RCDD (Registered Communications Distribution Designer) certification.
Building Industries Consulting Service International
10500 University Center Drive, Ste 100
Tampa, FL 33612-6415
Tel: (813) 979-1991, 1-800-BICSI-05
Fax: (813) 971-4311

CABA
Canadian Automated Buildings Association
M-20, 1200 Montreal Rd
Ottawa, ON K1A 0R6
Tel: (613) 990-7407
Fax: (613) 954-5984

Canadian Regulatory Agency
www.crtc.gc.ca
CRTC Public Affairs
Ottawa, K1A 0N2
819-997-0313

CSA
Canadian Standards Association
178 Rexdale Blvd
Rexdale, Ont
Canada M9W 1R3
Tel: (416) 747-4000, Documents Orders: (416) 747-4044
Fax: (416) 747-2475

EIA/TIA
EIA and TIA documents may be purchased through Global Engineering
Documents at 1-800-854-7179
EIA
EIA Standards Sales Office
2001 Pennsylvania Ave., N.W.
Washington, DC 20006
Tel: (202) 457-4966

European Telecommunications Standards Institute (ETSI)
The European counterpart to ANSI
www.etsi.org

Federal Communications Commission
The FCC's National Call Center, which provides consumer information on
telephone-related issues, can be reached by calling 1-888-CALL-FCC
(1-888-225-5322)

For FCC documents contact the FCC's contractor for public records duplication:
Downtown Copy Center
1990 M Street, N.W., Suite 640
Washington, D.C. 20036
202-452-1422

Frame Relay Forum
The Frame Relay Forum is an association of vendors, carriers, users and consultants
http://www.frforum.com/

GED
Global Engineering Documents
1990 M Street W, Suite 400
Washington, DC 20036
Tel: (800) 854-7179 (CDN/USA)
(202) 429-2860 (International)
(714) 261-1455 (International)
Fax: (317) 352-8484

Global Engineering Documents (West Coast)
2805 McGaw Ave.
Irvine, CA 92714
800-854-7179

IEC
International Electrotechnical Commission
rue de Varembre, Case Postale 131,3
CH-1211
Geneva 20, Switzerland

IEEE
P.O. Box 1331
Pisctaway, NJ 08855

ISLUA
International SL-1 Users Association (find info)

ISO
International Organization for Standardization
1, rue de Varembre, Case Postale 56
CH-1211
Geneva 20, Switzerland
Tel: +41 22 34 12 40

ITU
International Telecommunications Union (ITU) -Telecommunications Standardization Sector, (formerly called CCITT)
International Telephone Union
Place des Nations
CH-1211
Geneva 20, Switzerland
http://www.itu.int/

NATD
National Association of Telecommunications Dealers
561-266-9440
PO Box 100
Hohokus, New Jersey 07423
201-444-8946
201-444-5113 Fax

National Electrical Code
For a copy of the National Electrical Code contact:
OPAMP Technical Books
1033 North Sycamore Avenue
Los Angeles, CA 90038
800-468-4322

NFPA (US National Electrical Code (NEC) and other docs)
National Fire Protection Association
One Battery March Park, P.O. Box 9146
Quincy, MA 02269-9959
Tel: (800) 344-3555
Fax: (617) 984-7057

NIST
U.S. Dept. of Commerce
National Institute of Standards and Technology
Technology Building 225
Gaithersburg, MD 20899

NIUF
North American ISDN Users Forum
NIUF Secretariat
National Institute of Standards and Technology
Bldg 223, Room B364
Gaithersburg, MD 20899
Tel: (301) 975-2937
Fax: (301) 926-9675
Internet: sara@isdn.ncsl.nist.gov

NRC of Canada
Client Services
Institute for Research in Construction
National Research Council of Canada
Ottawa, ON K1A 0R6
Tel: (613) 993-2463
Fax: (613) 952-7673

NRUG
National Rolm Users Group
401 North Michigan Avenue
Chicago, IL 60611-4267
312-321-6804

NTIS
U.S. Dept. of Commerce
National Technical Information Service
5285 Port Royal Rd
Springfield, VA 22161
Tel: (703) 487-4650 / (800) 336-4700 (rush orders)
Fax: (703) 321-8547

SCC
Standards Council of Canada
1200-45 O/Connor St
Ottawa, Ont Canada K1P 6N7
Tel: (613) 238-3222
Fax: (613) 995-4564

TCA
Tele-Communications Assocation
701 N. Haven Avenue
Suite 200
Ontario, CA 91764-4925
909-945-1122

Telecommunications Research and Action Center (TRAC)
TRAC is a non-profit consumer organization devoted to educating consumers
on their telecommunication choices. You can request a publications list by
sending a stamped, self-addressed envelope to:
TRAC
P.O. Box 27279
Washington, DC 20005

Tele-Consumer Hotline
The Tele-Consumer Hotline us an impartial and independent consumer edu-
cation service that offers free publications to address your telecommunica-
tions-related concerns and issues.
For free publications, send a self-addressed stamped envelope to:
Tele-Consumer Hotline
P.O. Box 27207
Washington, DC 20005.

TERC

Telecommunications Equipment Remarketing Council
2000 M Street, N.W., Suite 550
Washing, D.C. 20036-3367
800-538-6282; 202-296-9800

TIA

Telecommunications Industries Association (TIA)
2500 Wilson Boulevard, Suite 300,
Arlington, VA 22201
Tel: (703) 907-7700
Fax: (703) 907-7727

UL

Underwriters Labs Inc.
333 Pfingsten Road,
Northbrook, Illinois 60062-2096 USA
Tel: (800) 676-9473 (from CDN/USA East coast)
(800) 786-9473 (from CDN/USA West coast)
(708) 272-8800 (International)
Fax: (708) 272-8129
0002543343@mcimail.com
MCI Mail: 254-3343

Glossary

900 number: 900 is an area code accessible only within the United State and Canada. Calls to this area code are billed to the caller at either a flat rate per call or a fixed amount per minute. The long distance carrier collects these charges from the caller and forwards a portion of them to the company (merchant) being called. A long distance carrier will connect a 900 number only by a T-1 circuit. If a T-1 circuit is not desirable, the 900 number will be connected through a service bureau. 900 numbers are often used to provide entertainment or support services.

976 number: Within most area codes in the United States and Canada, a special prefix, 976, is available and charges callers a fixed rate per minute. A portion of the fees collected by the telephone company are forwarded to the company or merchant being called. This is similar to a 900 number except the call is available only within a certain area code. (Callers outside the area code can access the number, but they must pay long distance charges in addition to the per-minute charge for the service.) This service is offered by the local telephone company, not by a long distance carrier. There are additional restrictions on content and rates are typically limited to less than $1 per minute.

A

abbreviated dialing: A feature that permits the calling party to dial the destination telephone number in fewer than the normal number of digits. Abbreviated dialing numbers must be set up in advance of their use. Speed dialing is a typical example of abbreviated dialing.

access: The method, time, circuit, or facility used to enter the network.

access coordination: The design, ordering, installation, preservice testing, turn-up and maintenance on local access services.

access line: The circuit used to enter the communications network.

account codes: Also known as Project Codes or Bill-Back Codes. Account codes are additional digits dialed by the calling party that provide information about the call. Typically used by hourly professionals (accountants, lawyers, etc.) to track and bill clients, projects.

ACD: (Automatic Call Distributor) A specialized phone system designed originally to evenly distribute heavy incoming calls, now increasingly used by companies also making outgoing calls. An ACD performs several functions. 1) It recognizes and answers incoming calls. 2) It looks in its database for instructions about what to do with a call. 3) According to the instructions, it will send the call to a recording site or forward it to an agent if the primary target user is online, or 4) record a message after the caller has heard the canned instructions.

ACNA: (Access Carrier Name Abbreviation) For example, WorldCom's ACNA is "WTL." There can be multiple carrier identification codes (CICs) per ACNA.

adapter: A physical device that allows one hardware or electronic interface to be adapted (accommodated without loss of function) to another hardware or electronic interface. In a computer, an adapter is often built into a card that can be inserted into a slot on the computer's motherboard. The card adapts information that is exchanged between the computer's microprocessor and the devices that the card supports.

ADC: (Analog to Digital Converter)

address translation: The process of converting external addresses into standardized network addresses and vice versa. Facilities interconnection of multiple networks which each have their own address plan.

ADSL: (Asymmetric Digital Subscriber Line) Asymmetric refers to the flow of data, meaning that more data is sent in one direction than the other. The technology is ideal for applications such as Internet access, remote client/server links and video, because more bandwidth is used sending data downstream to the customer than upstream to the server.

agent: A person or an organization that acts on behalf of another. In the telecommunications industry, agents typically are 1) independent individuals or companies that market the services of a carrier as if they were employees of that carrier, 2) a person in a telephone center who answers the call, answers questions, takes orders, etc.

aggregate discount: A discount applied to multiple services based on the total dollar value of those services.

aggregator: An independent entity that brings several subscribers together to form a group that can obtain long-distance service at a reduced rate. Subscribers are billed by the original interexchange carriers (IXC). The aggregator, different from a reseller, only provides the initial set-up of the plan.

AIN: (Advanced Intelligent Network) A dynamic database used in Signaling System 7. It supports advanced features by dynamically processing the call based upon trigger points throughout the call handling process and feature components defined for the originating or terminating number. AIN is able to query LNP (Local Number Portability) databases.

AIOD: (Automatic Identified Outward Dialing) An option on a PBX that specifies the extension number, instead of the PBX number on outward calls (for internal billing).

alternate access: A form of local access where the provider is not the LEC, but is authorized or permitted to provide such service.

alternate access carriers: Local exchange carriers in direct competition with the RBOCs. Until recently found only in the larger metropolitan areas. Examples: Teleport and Metropolitan Fiber Systems.

alternative operator services: Operator services provided by a company other than a LEC, RBOC or AT&T that is authorized to provide such service.

AMA record: (Automatic Message Accounting) See CDR.

AMI: One of two protocols used by a T-1 circuit for interfacing between a long distance carrier's switch and a customer's equipment. The other protocol is B8ZS (which see). A T-1 interface board must be configured for the correct protocol to work properly.

ampere: Unit of electric current strength equal to the flow of one coulomb per second.

analog station port: An extension phone interface on a business telephone system that provides and responds to standard Bell-system analog signals such as dial tone, flash hook, and touch tones.

analog: Comes from the word "analogous," which means "similar to." In telephone transmission, the signal being transmitted, voice, video, or image, is "analogous" to the original signal. But in telecommunications, analog means telephone transmission and/or switching which is not digital. The human voice is an analog signal. Analog technology refers to electronic transmission accomplished by adding signals of varying frequency or amplitude to carrier waves of a given frequency of alternating electromagnetic current. Broadcast and phone transmission have conventionally used analog technology. Analog also connotes any fluctuating, evolving, or continually changing process. Analog is usually represented as a series of sine waves. The term originated because the modulation of the carrier wave is analogous to the fluctuations of the voice itself. A modem is used to convert the digital information (a series of 1s or 0s in specific patterns) in computers to analog signals for phone lines and to convert analog phone signals to digital information for the computer. When voice signals are translated to digital signals the analog "wave" is sampled, meaning the signal is recorded at specific intervals. Each sample of the amplitude of analog waves is translated into a digital representation and then transmitted. The more frequent the sampling, the better the analog wave will be recreated at the remote end of the transmission.

ancillary features: Subordinate, supplementary, subcomponent characteristics and capabilities that are marketing options of products and services.

ANI: (Automatic Number Identification) 1) The number associated with the telephone station(s) from which switched calls are originated (or terminated). 2) A software feature associated with Feature Group D (and optional on Feature Group B) circuits. ANI provides the originating local telephone number of the calling party. This information is transmitted as part of the digit stream in the signaling protocol, and included in the call detail record for billing purposes. 3) ANI may also be used to refer to any phone number.

ANSI: (American National Standards Institute) A United States-based organization which develops standards and defines interfaces for telecommunications.

answer supervision: The off-hook indication sent back to the originating end when the called station answers.

area code: A 3-digit code designating a toll center in the US and Canada. The 1995 adoption of the North American Numbering Plan (NANP) expanded available numbers from 152 to 792.

area code routing: Route calls based on the originating ANI NPA (area code). See NPA-NXX Routing.

area of service: (AOS) The geographical area supported by a communication service. For 800 numbers, if AOS is "CC," it is using Complex Call routing.

ARP: (Address Resolution Protocol) under TCP/IP. Used to dynamically bind a high level IP address to a low-level physical hardware address. ARP is limited to a single physical network that supports hardware broadcasting.

ARS: (Automatic Route Selection) The phone system automatically chooses the least costly way of sending a long distance call.

ASR: (Access Service Request) A document (or data transaction) sent to the LEC to order the local access portion of a circuit.

ASCII: (American Standard Code for Information Exchange) A popular code used in small computers for changing letters and numbers into the digital code (zeros and ones), used by all computers.

asymmetric: A term used in high speed transmission to denote a greater flow of data in one direction than the other. (see ADSL)

asymmetric digital subscriber line: (see ADSL)

asynchronous: (not synchronous) A form of concurrent input and output communication transmission with no timing relationship between the two signals. Slower-speed asynchronous transmission requires start and stop bits to avoid a dependency on timing clocks (10 bits to send on 8-bit byte). (Contrast with Synchronous)

asynchronous transfer mode: (ATM) An international ISDN high-speed, high-volume, packet-switching transmission protocol standard. ATM uses short, uniform, 53-byte cells to divide data into efficient, manageable packets for switching through a high-performance communications network. The 53-byte cells contain 5-byte destination address headers and 48 data bytes. ATM is the first packet-switched technology designed from the ground up to support integrated voice, video, and data communication applications. It is well-suited to high-speed WAN transmission bursts. ATM currently accommodates transmission speeds from 64 Kbps to 622 Mbps, and may support gigabit speeds in the future. (See cell relay.) Because ATM is a switched service, a connection has to be established between the sender and the receiver before data can be sent.

asynchronous transfer mode adaptation layer: (AAL) A series of protocols enabling ATM to be made compatible with virtually all of the commonly used standards for voice, data, image and video.

ATIS - Alliance for Telecommunications Industry Solutions, formerly the Exchange Carriers Standards Association (ECSA), was created in 1983. This is an association comprised of telecommunications companies involved with all aspects of communications, including manufacturers and vendors. Various subcommittees to ATIS exist. Subcommittees of specific interest to this document are the CLC and the T1-Telecommunications Committee under ANSI.

ATM: (Asynchronous Transfer Mode) See above.

attenuation: A loss of signal strength in a lightwave, electrical or radio signal usually related to the distance the signal must travel (example: fiber optic transmission must be regenerated approx. every 30 miles). Fiber optic attenuation is caused by transparency of the fiber, a too-tight bend of the fiber, nicks in the fiber, splices, poor fiber terminals, FOTs, etc. Electrical attenuation is caused by the resistance of the conductor, poor (corroded) connections, poor shielding, induction, RFI, etc. Radio signal attenuation may be due to atmospheric conditions, sun spots, antenna design / positioning, obstacles, etc.

auth code: (Authorization Code) A number used for security purposes to gain access to an interexchange carrier's network. Authorization codes are inherently required for all feature Group-A and feature Group- B circuits without ANI reporting. Authorization codes are also required for travel service and cut-through capabilities on feature Group-D circuits.

automated attendant: Usually provided with most voice mail systems, plays a recorded greeting, then transfers a call to the extension or phone line the caller has selected.

automatic number identification: (See ANI.)

automatic ring down: (ARD) A private line connecting a station instrument in one location to a station instrument in a distant location with automatic two-way signaling. The automatic two-way signaling used on these circuits causes the station instrument on one end of the circuit to ring when the station instrument on the other end goes off-hook. This circuit is sometimes called a "hot-line" because urgent communications are typically associated with this service. ARD circuits are commonly used in the financial industry. May also have one way signaling. Station "A" rings Station "B" when Station "A" goes off hook, but Station "B" cannot ring Station "A."

B

B channel: A component of ISDN interfaces. B channels can carry 64 Kbps in both directions, either voice or data.

B8ZS: (Bipolar with eight Zero Substitution) A protocol used by a T-1 circuit for interfacing between a long distance carrier's switch and a customer's equipment. A T-1 interface board must be configured for the correct protocol to work properly. More technically, a clear channel line coding option on DS-1 service allows the DS-1 user to obtain greater throughput and functionality from their DS-1 facilities. The use of B8ZS allows users to transmit data at a rate of 64 Kbps per DS-0, achieving what is referred to as a clear channel. Applied against all 24 DS-Os on a DS-1, the effective data throughput of the DS-1 facility is increased with B8ZS from 1.344 Mbps to 1.536 Mbps, a 14% increase in throughput. CSUs with B8ZS support are required on both ends of the user's circuit.

Baby Bells: See RBOC.

backbone: Network of broadband connections between switches.

background music: Music played through speakers in the ceiling and/or through speakers in each telephone, throughout the office.

backup: A copy of computer data on an external storage medium, such as a floppy disk or tape or additional computers and telephone systems that can be used if the primary system fails.

ballot: A release form that authorizes a customer's long-distance phone service to be switched to (another) long-distance carrier, or reseller.

BAN: (Billing Account Number) Used by telephone companies to designate a billing account, a customer or customer location that receives a bill. A customer may have any number of BANs.

banded rates: Tariffed rates which may be changed by the carrier within a specified range. Frequently, state commissions require notice to the commission prior to each change. Banded rates are being used less frequently today.

bandwidth: A measure of the communication capacity or data transmission rate of a circuit. The total frequency spectrum (in Hertz, cycles per second) that is allocated or available to a channel, or the amount of data that can be carried (in bps) by a channel. Bandwidth is used to mean 1) how fast data flows on a given transmission path, and more technically, 2) the width of the range of frequencies on which electronic signals are carried on a given transmission medium. Any digital or analog signal has a bandwidth. Generally speaking, bandwidth is directly proportional to the amount of data transmitted or received per unit time. In a qualitative sense, bandwidth is proportional to the complexity of the data for a given level of system performance. For example, it takes more bandwidth to download a photograph in one second than it takes to download a page of text in one second. Large sound files, computer programs, and animated videos require still more bandwidth for acceptable system performance. Virtual reality (VR) and full-length three-dimensional audio/visual presentations require the most bandwidth. In digital systems, bandwidth is proportional to the data speed in bits per second (bps). Thus, a modem that works at 57,600 bps has twice the bandwidth of a modem that works at 28,800 bps. In analog systems, bandwidth is defined in terms of the difference between the highest-frequency signal component and the lowest-frequency signal component. A typical voice signal has a bandwidth of approximately three kilohertz (3 kHz); an analog television (TV) broadcast video signal has a bandwidth of six megahertz (6 MHz) — some 2,000 times as wide as the voice signal. Communications engineers once strove to minimize the bandwidths of all signals, while maintaining a minimum acceptable level of system performance. This was done for at least two reasons: 1) low-bandwidth signals are less susceptible to noise interference than high-bandwidth signals; and 2) low-bandwidth signals allow for a greater number of communications exchanges to take place within a specified band of frequencies. However, this simple rule no longer applies generally. For example, in spread-spectrum communications, the bandwidths of signals are deliberately expanded. In digital cable and fiber optic systems, the demand for ever-increasing data speeds outweighs the need for bandwidth conservation. In the electromagnetic radiation spectrum, there is only so much available bandwidth to go around, but in hard-wired systems, available bandwidth can literally be constructed without limit by installing more and more cables.

base rate: The nondiscounted "per minute" charge for measured service.

basic rate interface: (BRI) ISDN offering that allows 2 64 Kbps and 1 16 Kbps channels to be carried over a typical single pair of copper wires. Through the use of BONDING (Bandwidth on Demand) the two 64 Kbps channels can be combined to create more bandwidth as it becomes necessary.

battery backup (system backup): A battery which provides power to your phone system, network equipment, or other important equipment when the main AC power fails especially during blackouts and brownouts.

baud: The older term being replaced by bps, bits per second. The number of signaling elements that can be transmitted per second on a circuit. e.g. When a modem is used to send digital information on an analog line, baud refers to the speed that the circuit can change from the tone used to represent a binary zero to the tone used to represent a binary one (or vice versa). In an average data stream, one baud is roughly equivalent to one bit per second on a digital transmission circuit.

B-channel: (Bearer channel) In the Integrated Services Digital Network (ISDN), the B-channel is the channel that carries the main data. In ISDN, there are two levels of service: the basic rate (see BRI), intended for the home and small enterprise, and the primary rate, for larger users. Both rates include a number of B (bearer) channels that carry data, voice, and other services, and a D (delta) channel that carries control and signaling information. The basic rate consists of two 64 Kbps B-channels and one 16 Kbps D-channel. Thus, a basic rate user can have up to 128 Kbps service. The primary rate consists of 23 B- channels and one 64 Kpbs D-channel in the United States, or 30 B-channels and 1 D-channel in Europe.

Bell customer code: A three-digit numeric code, appended to the end of the main billing telephone number, that is used by local exchange carriers to provide unique identification of customers.

Bell operating company: (BOC) The local (or regional) telephone company that owns and operates lines to customer locations and Class 5 central office switches. BOCs have connections to other central offices (COs), tandem (Class 4 Toll) offices, and may connect directly to InterExchange carriers (IECs) like WorldCom, AT&T, MCI, and Sprint. BOC may refer to the nineteen Bell operating companies that are owned by the seven regional holding companies (RHCs), not including Cincinnati Bell or Southern New England Telephone. The BOC role was originally defined by the 1982 Modified Final Judgement that specified the terms of the AT&T divestiture). For example, the three BOCs, Mountain Bell, Northwestern Bell, and Pacific Northwest Bell are owned by the U.S. West RHC. Each BOC may service more than one local access and transport area (LATA), but BOCs are generally constrained from providing long distance service between LATAs.

BER: See Bit Error Rate

beta test: A secondary product test performed by a selected set of "early support" end user(s) or customer(s) (under special contract) prior to the general availability of the product.

BICSI: (Building Industries Consulting Service International) A non-profit professional association for those engaged in voice\data cable plant design and installation. Administers the Registered Communications Distribution Designer (RCDD) and Local Area Network (LAN) Specialist certifications and provides related training.

billing account number: (BAN) Used by telephone companies to designate a customer or customer location that will be billed. A single customer may have multiple billing accounts.

bill-to-room: A billing option associated with operator assisted calls that allows the calling party to bill a call to their hotel room. With this option, the carrier is required to notify the hotel, upon completion of the call, of the time and charges.

binary file transfer: (BFT) The transmission of binary files between communicating devices.

Bird: Satellite (informal slang)

BISDN (or B-ISDN): (Broadband Integrated Services Digital Network) See ISDN. A packet switching technique which uses packets of fixed length, resulting in lower processing and higher speeds. Also see ATM or Cell Relay bit error rate: (BER) The rate at which errors occur in a stream of transmitted data. The BER may be expressed in terms of a percentage of error-free seconds or as a percentage of error-free bits.

bit: (BInary digiT) The smallest amount of information that can be transmitted. In binary digital transmission, a bit has a single binary value, either 0 or

1. A combination of bits can indicate an alphabetic character, a numeric digit, or perform a signaling, switching or other function. Although computers usually provide instructions that can test and manipulate bits, they generally are designed to store data and execute instructions in bit multiples called bytes. In most computer systems, there are eight bits in a byte. The value of a bit is usually stored as either above or below a designated level of electrical charge in a single capacitor within a memory device. Half a byte (four bits) is called a nibble. In some systems, the term octet is used for an eight-bit unit instead of byte. In many systems, four eight-bit bytes or octets form a 32-bit word. In such systems, instruction lengths are sometimes expressed as full-word (32 bits in length) or half-word (16 bits in length).

block calls: Prevent calls from completing to destination. May be by customer request (block calls from or to certain NPAs, NXXs, States, LATAs), or inadvertent due to network problems of outage, overload, etc.

BOC: See Bell Operating Company

bong: An interactive signal that prompts the originating end user to enter additional information.

bps: (Bits Per Second) Upper/lower case of the "b" in the acronym is significant. In data communications, bps is a common measure of data speed for computer modems and transmission carriers. As the term implies, the speed in bps is equal to the number of bits transmitted or received each second. The duration d of a data bit, in seconds s, is inversely proportional to the digital transmission speed s in bps: $d = 1/s$. Larger units are sometimes used to denote high data speeds. One kilobit per second (Kbps) is equal to 1000 bps. One megabit per second (Mbps) is equal to 1,000,000 bps or 1000 Kbps. Computer modems for twisted-pair telephone lines usually operate at speeds between 14.4 and 57.6 Kbps. The most common speeds are 28.8 and 33.6 Kbps. So-called "cable modems," designed for use with TV cable networks, can operate at more than 100 Kbps. Fiber optic modems are the fastest of all; they can send and receive data at many Mbps. The bandwidth of a signal depends on the speed in bps. With some exceptions, the higher the bps number, the greater is the nominal signal bandwidth. Speed and bandwidth are, however, not the same thing. Bandwidth is measured in standard frequency units of kilohertz (khz) or megahertz (MHz). Data speed is sometimes specified in terms of baud, which is a measure of the number of times a digital signal changes state in one second. Baud, sometimes called the "baud rate," is almost always a lower figure than bps for a given digital signal. The terms are often used interchangeably, even though they do not refer to the same thing. If a computer modem is said to function at "33,600 baud" or "33.6 kilobaud," the term is probably being misused, and the numbers actually indicate bps.

Bps: Bps (8-bit) bytes per second

BRI: (Basic Rate Interface in ISDN) 3 digital signals over a single pair of copper wires: 2 voice (B) channels and 1 signaling (D) channel which allow voice and fax on a single pair of wires. In the Integrated Services Digital Network

(ISDN), there are two levels of service: the BRI, intended for the home and small enterprise, and the Primary Rate Interface (PRI), for larger users. Both rates include a number of B (bearer) channels and a D (delta) channel. The B channels carry data, voice, and other services. The D channel carries control and signaling information. The BRI consists of two 64 Kbps B channels and one 16 Kbps D channel. Thus, a Basic Rate user can have up to 128 Kbps service. The PRI consists of 23 B channels and one 64 Kbps D channel in the United States or 30 B channels and 1 D channel in Europe. The typical cost for Basic Rate usage in a city like Kingston, New York is about $125 for phone company installation, $300 for the ISDN adapter, and an extra $20 a month for a line that supports ISDN. For more information, see ISDN.

bridge: A local area network (LAN) internetworking device that filters and passes data between LANs based on Layer 2 (Medium/Media Access Control [MAC] layer) information. Bridges do not use any routing algorithms. (Compare router. Contrast gateway - dissimilar protocols.)

broadband: Broadband refers to a high-capacity telecommunication circuit/path that provides multiple channels of data over a single communications medium using frequency division multiplexing. It usually implies a speed greater than 1.544 Mbps. (Contrast with wideband and narrowband)

brouter: A term used by some vendors, normally referring to a bridge also having some of the characteristics of a router.

BT: (Busy Tone) which see.

BTN: (Billing Telephone Number) The phone number associated, for billing purposes, with the working phone number. (See BAN.)

burst rate: Top speed (largest bandwidth) normally needed by a user of a frame relay system.

burst tolerance: (BT) Different levels of BT are among descriptors of classes of service. ATM switches must provide buffering to absorb bursts.

burst transmission: (BT) Transmission of high volumes of data in short periods of time.

bursty (or batchy): Communications characterized by high volumes of data transmitted intermittently, as opposed to steady-stream data.

busy tone: (BT) The normal line-is-busy or off-the-hook tone is sounded once a second. The trouble-on-the-line busy tone is sounded twice each second.

butt-set, butt-in, buttinski: Hand-carried test telephone used to monitor, dial, and talk on conventional analog telephone lines. So named because the technician can clip onto a pair and "butt in" to a conversation.

bypass: Access an interexchange carrier (IEC) other than the customer's equal access carrier by dialing 10+CIC Code. (For example, Bypass to WorldCom by dialing "1010555"). See Walkthrough, CIC Code.

bypass service: The use of facilities other than those of the local exchange carrier (LEC), a facilities bypass, or the use of operating telephone company private lines, a service bypass, to connect a customer location to a point of presence (POP, the local long distance carrier) or to another customer location.

byte: In most computer systems, a byte is a unit of information that is eight bits long. A byte is the unit most computers use to represent a character such as a letter, number, or typographic symbol (for example, "g", "5", or "?"). A byte can also hold a string of bits that need to be used in some larger unit for application purposes (for example, the stream of bits that constitute a visual image for a program that displays images). In some computer systems, four bytes constitute a word, a unit that a computer processor can be designed to handle efficiently as it reads and processes each instruction. Some computer processors can handle two-byte or single-byte instructions. A byte is abbreviated with a "B". (A bit is abbreviated with a small "b".) Computer storage is usually measured in byte multiples (for example, an 820 MB hard drive holds a nominal 820 million bytes (megabytes) of information. (The number is actually somewhat larger since byte multiples are calculated in powers of 2 and we express them as decimal numbers .)

C

cable cut: Service outage caused by a cut or damaged cable.

CABS: (Carrier Access Billing System) CABS processes records that are needed for Carrier Access Billing. It calculates rates/charges based on applicable tariffs, posts the resulting charges to the carriers' accounts, and bills the carriers on a regular basis.

CAC: 1) (Carrier Access Code, which see) 2) (Connection Admission Control, which see)

cadences: distinctive ring patterns available as options for a phone system.

call accounting system: 1) A call accounting system is used to record information about telephone calls, organize that information, and upon being asked, prepare reports, printed or to disk. 2) More technically, a database-management system that accepts CDR/SMDR records, analyzes them, applies costing data, and produces reports that assist in tracking telecommunications charges.

call back: A feature of some voice and data telephone systems. If the number called is busy, the button or code for "call-back" can be pushed and when the phone is free, the phone system will call you back and simultaneously call the number originally dialed.

call blocking: An option that prevents specified incoming calls.

call detail record: (CDR) An accounting record produced by switches to track call type, time, duration, facilities used, originator, destination, etc. CDRs are used for customer billing, rate determination, network monitoring, and facility capacity planning. CDRs represent unrated calls (to be processed by rating) in contrast to toll calls, which are rated calls. (See SMDR.)

call duration: The period of time that begins with answer supervision (destination off hook) and ends when the call is terminated.

call forwarding: A service available in many central offices, and a feature of many PBXs and some hybrid PBX/key systems, which allows an incoming call to be re-directed from one phone number or extension to another under various conditions. Call Forwarding Variable (CFV) is used to describe call forwarding that can be set by the user. CFV requires the user to physically dial a code and a destination number, then sends the call from the first trunk or extension immediately to the specified number. Call Forward No Answer routes calls to a second number when the first number is not answered after a preset number of rings. Call Forward Busy redirects the calls from a busy trunk or extension to the designated target. These terms originally applied to Centrex service.

call hold: A phone or PBX feature that allows a call to be placed on hold.

call park: Similar to placing a call on hold, but the line must be accessed by one or 2 digits. Use of the digits allows picking up the call from any phone in the system.

call pickup: A phone is ringing but not yours. With call pickup, you can punch in a button or two on your phone and answer that person's ringing phone.

call progress: The ability of a voice mail card to determine the results of a call transfer by listening for signals sent by the phone company such as busy signal, dial tone, ring, etc.

call queuing: The process of placing calls in a line to wait for a busy extension to become available. A voice mail system which supports call queuing will typically play a sound file to the caller while they are waiting and it will periodically attempt to complete the call transfer. Callers usually can transfer out of the call queue and instead leave a voice mail message.

call routing tree (or call tree): A graphical display of complex call routing decision logic or computer function call sequence. Documents function usage. Used for change impact analysis.

call transfer: The process of automatically moving a telephone call from one extension or phone number to a different extension or phone number. PBX or KSU telephone systems must be used, or a service called Centrex or Three-Way Calling must be installed on the telephone line.

call transfer, blind: When a call is transferred by a voice mail system, the voice mail system does not monitor the transfer in any way. It simply issues the transfer commands.

call transfer, semi-blind: Voice mail system listens only for a busy signal. If a busy signal is received, the voice mail will reverse the transfer and can take a message. If any other sound or signal is received, the voice mail simply completes the transfer.

call transfer, supervised: Voice mail system monitors the complete transfer process, listening for busy signals and ring signals. If the extension is not answered or is busy, the voice mail system will reverse the transfer.

call tree: (see call routing tree)

call type: Identification of call type as 1+, 0+, 800, etc.

call waiting: A feature of phone systems that informs of incoming calls by a beep, a light, or a message on the screen.

called station: Also known as called party destination node on the network. The telephone number to which a call is directed or terminated.

caller ID: A service available in some parts of the country which will provide the caller's phone number to the person being called. Special equipment is required to receive this information.

calling card: A telecommunication credit card with an AuthCode for using a long distance carrier when the customer is away from home or office, or automatic number identification (ANI).

calling station: Also known as calling party or origination node. The fax machine transmitting the message.

calling station identification: (CSID) The usual fax name/number field printed across the top of the fax.

CAP: (Competitive Access Provider) An alternative, competitive local exchange carrier. A telecommunications carrier that provides access services which are alternate to (or which bypass) a local exchange carrier. Also referred to as AAP - Alternate Access Provider.

CAP: (Carrierless Amplitude and Phase) A very efficient coding technique that modulates transmit and receive signals into 2 wide-frequency bands and supports several of the DSL variations as well as working with DMT.

capture buffer: Also called a "data recorder" or "buffer box," this is a task-dedicated device that captures CDR/SMDR (which see) records produced by a PBX, retaining them until "polled" by a computer running call accounting or telemanagement software. Capture buffers store data in nonvolatile RAM, so it is safe from power-outages. They let you separate the ongoing task of data-collection from the periodic need for data-analysis saving you from having to dedicate a general-purpose PC to the data-capture function. Remote-pollable capture buffers let a single site collect and manage call data from multiple locations and are frequently used by call accounting service bureaus.

card issuer identifier code: (CIID, pronounced "sid") A code issued with certain calling cards. AT&T's CIID cards cannot be used by other interexchange carriers but can be used by local exchange carriers (LECs).

carrier access code: (CAC) also called Carrier Identification Code (CIC) The 3-digit numbers customers use to reach interexchange carriers. That carrier's identification number. The primary carrier is reached by pressing "1" and then

411

the area code and 7-digit customer number. Secondary carriers can be reached by pressing "10" plus the CAC (CIC) of the secondary carrier.

carrier circuit: A higher level circuit (DS-1, DS-3, Transmission System, etc.) that has been designed to carry lower-level circuits (DS-0, DS-1).

carrier facility assignment: (CFA) An identifier for the Telco network point where an interexchange carrier (IEC) connects.

carrier identification code: (CIC) A three digit number used with Feature Groups B and D to access a particular IEC's switched services from a local exchange line. One or more CIC codes are assigned to each carrier. There may be multiple CICs per access carrier. (See Bypass)

carrier split: Use of 800 service call routing features to divide 800 calls between two or more IECs. Split may be by % allocation, origination NPA, time of day, etc.

carrier: A telecommunications provider that owns switching equipment.

carrierless amplitude and phase: See CAP for description.

CAS: (Communicating Applications Standard (or Specification), or Centralized Attendant Service, which see for descriptions.

casual calling: Allows any automatic number identification (including undefined ANIs) to access a given carrier. For example, if the originator is calling from a non-coin phone, dialing 10555+destination number will route the call through WorldCom and it will be billed to the originating phone number.

casual customer: Any person or organization that dials any CIC Code. (Not necessary to presubscribe to the carrier.)

CAT 3: (Category 3) Cabling and cabling components designed and tested to transmit cleanly 16 megahertz of communications. Used for voice and data/LAN traffic to 10 megabits per second. The most common cabling used for voice telephone. Specs are defined by the FCC.

CAT 5: (Category 5) Cabling and cabling components designed and tested to transmit cleanly 100 megahertz of communications. Used for voice and data/LAN traffic to 155 megabits per second.

CATV: (Cable Television - Community Antenna Television) A community television system, served by cable and connected to a common (set of) antenna(s). 1994 Federal legislation may allow them to compete with local exchange carriers (LECs) for telephone service (on the Information Superhighway).

CBR: (Constant Bit Rate) Defined by the ATM Forum as, "an ATM service which supports a constant or guaranteed rate to transport services such as video or voice as well as circuit emulation which requires rigorous timing control and performance parameters."

CBUD: (Call Before U Dig) Operational management system for protection of fiber facilities. May have electronic geographic maps of states, counties, and city streets where the carrier has buried facilities, upon which reported construction activities are automatically mapped. Human technicians verify that the activities do not pose a danger to the facilities. Technicians may be dispatched to the construction site when facilities may be at risk.

CCITT: (Comite Consultatif Internationale de Telegraphique et Telephonique, or Consultant Committee on International Telephone and Telegraph) An international organization which develops standards and defines interfaces for telecommunications (now known as International Telecommunications Union-Telecommunications, ITU-T). Located in Geneva, Switzerland, ITU-T is the primary international body for fostering cooperative standards for telecommunications equipment and systems.

CCS: 1) Common channel signaling one hundred (Roman Numeral C) Calling Seconds 2) A standard unit of traffic, used in communications engineering. (See Erlang.)

CDR: (Call-Detail Record) A line of ASCII text, output by a business phone system (usually on a serial port) in response to calling activity (outbound, inbound, transfers). A typical record contains several "fields," showing the extension making a call, the called number, call duration (or call start- and end-times), and the trunk used. Special-format records may also be produced when the switch flips from Day mode to Night mode, when somebody accesses the maintenance port, or when other periodic changes occur. There is no industry-standard format for call detail records - each PBX manufacturer has its own format for presenting the data. Analysis of call-detail records reveals the cost of phone usage (for department/project accounting, comparison with carrier bills, and other applications), points up incidents of toll fraud and telabuse, and permits assessment of provisioning (for example, Are all the trunks being used?).

CDSL: (Consumer Digital Subscriber Line) CDSL is a trademarked version of xDSL. It is slower than ADSL but, similar to DSL Lite, does not require a splitter at the user end.

cell: 1) Packet switching information grouped in units of uniform size. Cells are fixed-length packets (ATM 53-byte cells). 2) A small group acting as a unit in a larger organization (for example, one of the separate geographical areas covered by a radio transceiver antenna in a multi-antenna cellular phone system.

cell relay: Packet switching technique which uses packets of fixed length, resulting in lower processing speeds. Also known as BISDN and ATM.

cellular service type: Type 1 - ANI only identifies the mobile cellular system, Type 2 - ANI identifies the mobile directory number (DN) placing the call, but does not necessarily identify the true call point of origin.

central office: (CO) 1)Telephone company facility where subscribers' lines are joined to switching equipment for connecting other subscribers to each other, locally and long distance. 2) More technically, one local Class 5 switch with lines to customer locations. (Usually fewer than 100,000 telephone lines per central office.) COs are usually owned and operated by local exchange carriers (LECs) or BOCs. COs have connections to tandem (Class 4 Toll) offices, and often connect directly to other COs and interexchange carriers (IECs) like WorldCom, AT&T, MCI, and Sprint. A CO is a major equipment center designed to serve the communications traffic of a specific geographic area. CO coordinates are used in mileage calculations for local and interexchange service rates. A non-conforming CO is one that does not (yet) support equal access.

central office trunk: 1) A trunk between central offices. 2) A trunk between public and private switches. 3) Physically, the 2-wire electrical path between the phone company central office and the customer.

centralized attendant service: (CAS) A group of switchboard operators answering all incoming calls for several locations within a geographic area.

CentraNet: See Centrex.

centrex: A service that is functionally similar to a customer-premise PBX, but provided by means of equipment located in a central office by a local Bell or GTE telephone company. This service provides call transfer functions to any other telephone number, even numbers in a different city or country. Some types of call forwarding functions are also available. This service runs over standard analog telephone lines. Centrex is marketed under different names by different telephone companies. Some names include CentraNet and Plexar.

CEPT: (Conference on European Post and Telegraph) A European organization which develops standards and defines interfaces for telecommunications.

CFA: See Carrier Facility Assignment

CGA: (Carrier Group Alarm) A service alarm produced by a channel bank (which see) when an error in framing bits persists longer than the the (preset) interval allows.

CGI: (Common Gateway Interface) Scripts or programs on the Internet Web server that perform some action when a button or item on the monitor is "pushed."

channel: A telecommunications path (pipe) of a specific capacity (speed) between two locations in a network. (See DS-0 through DS-4.)

channel bank: A multiplexer that merges into and controls the flow of 24 voice and/or data circuits in a single, high speed digital transmission link (T1). Converts digital and analog as necessary. May not be required if both sending and receiving systems are digital (PBX). (See channel service unit).

channel extension/channel networking (service): Interfaces that allow high-speed computers to communicate with remote devices at local channel speeds (over T1/T3 lines).

channel service unit/data service unit: (CSU/DSU) Manages digital transmission, monitors signals for problems. Responds to central office commands. It performs many of the functions that modems do, but it does not have to convert digital signals to/from analog, because the end device and the underlying transmission facility are both digital.

channelize: To subdivide (or break out) a broadband transmission system into multiple communication channels.

CIC: See Carrier Identification Code (WorldCom = "555").

CIR: (Committed Information Rate) In a frame relay network, each premises visit charge (PVC) is assigned a committed information rate, measured in bits per second. The CIR represents the average capacity that the port connection should allocate to the PVC. This rate should be consistent with the expected average traffic volume between the two sites that the PVC connects. The CIR that is assigned to a PVC cannot exceed the speed of either the originating or terminating port connection.

circuit: 1) The physical connection (or path) of channels, conductors and equipment between two given points through which an electric current may be established. Includes both sending and receiving capabilities. 2) A switched or dedicated communications path with a specified bandwidth (transmission speed/capacity). 3) Physically, in telephony, a 4-wire 2-pair connection.

circuit switching: A switching method where a dedicated path is set up between the transmitter and receiver. The connection is transparent, meaning that the switches do not try to interpret the data.

city pair: Two cities between which an interexchange carrier (IEC) offers long-distance service. When ordering a new dedicated circuit or trunk group, "city pair" NPA/NXXs are used to determine the switch location.

Class 3: IEC, interexchange carrier. Hierarchical interconnection for Class 4 and optional Class 5 switches.

Class 4: Tandem office, toll office. Interconnection for Class 5 switches and long distance via Class 3 IECs. Optional direct connection to higher volume Class 4 sites. A Class 4 may also serve as a Class 5 CO.

Class 5: Central office, end office. Connection to local customer premises equipment and local switching. capacity typically is up to 100,000 lines, 1 to 10 area codes (see NXX).

class of service: (COS) 1)Each phone in a system may have a different collection of privileges and features assigned to it, such as access to WATS lines. Class of service assignments, if properly organized, is an important tool in controlling phone abuse. 2) A special limitation on what numbers can and cannot be called. International, 809, 809 + Canada, 48 contiguous states, etc.

class of service: (CoS as related to ATM QoS) (see QoS) Class 1, equivalent to digital private lines. Class 2, supports audio- and videoconferencing and multimedia. Class 3, handles protocols of synchronous connection-dependent systems. Class 4, supports connectionless (Internet) data systems.

client/server: (C/S) A distributed computing model in which clients request data and processing from servers. Servers usually have higher capacity than clients (but not necessarily). Client/server exploits less expensive hardware than host-based computing, but C/S application design and resource management must be more sophisticated. See Peer-To-Peer.

CLLI: (pronounced "silly") Common Location Language Identifier A unique identifier assigned to LEC end offices and tandem (Class 4 switch) toll offices groups. The CLLI code is the designation for a central office, or the area served by a CO. (CLLI is a BellCore standard) Example: "SNANTXFRCGO". Digits 5 & 6 are the state code, digits 7 & 8 are the CO name, digits 9 through 11 specify equipment type.

closed end: The end of a line (such as a WATS 800 or foreign exchange line) from which all calls are directed to or from a single point. Private lines normally have two closed ends.

CMSDB: (Call Management System Data Base) Service control point (SCP) for 800 Number Translation Database (To POTS).

CO: See Central Office.

COAM: (Customer Owned and Maintained Equipment)

COB: (Close Of Business, completed by end of business day)

COCOT: (Customer Owned Coin Operated Telephone)

CODEC: (enCOde/DECode) A device that converts (encodes) analog signals into a form for transmission on a digital circuit, a process known as Pulse Code Modulation (PCM). The digital signal is then decoded back to analog at the receiving end of the transmission link. CODECs allow voice and video transmission over digital links. CODECs may also support signal compression. (Contrast modem.)

coin phone: A coin-operated pay phone with restricted access to some services (for example, international calling). Coin phones have subclasses of public, semi-public, and private.

collect: A call that is paid for by the receiving/destination phone number. Requires approval/authorization of the person being called.

colocation: The placement of in-service customer telecommunications equipment at a carrier's central office, point of presence, or other network location.

common carrier: A carrier that holds itself out as serving the public (or a segment thereof) indifferently (that is, without regard to the identity of the customer and without undue discrimination). Common carriers may vary rates based on special considerations and may in fact serve only a small fraction of the general public.

common line/end user: A "common line" is the portion of the exchange carrier's facilities that extends from the customer's premises to the exchange carrier's end office. The NECA common line pool recovers the interstate portion of the costs of maintaining those facilities.

communicating applications standard (or specification): (CAS) Fax standard for fax and voice applications. A popular software interface to a fax board, an application programming interface (API) specification that supports programs sending data to other machines, including computers.

communication link: A system of hardware and software connecting two end users.

communications server: (comms server) Most communications servers are PC-based, but some made by phone system vendors are actually phone systems modified to act like PCs. The comms server "box" is a gateway between computers and the LAN. It translates back and forth from asynchronous signals to Lan signal packets, integrates PBX-type features, CT functions, and can run other signals alongside data over a PC network. PC based comms servers are expanded by connecting a PC network, usually using TDM (Time Division Multiplexing) across the communications buses of the PCs.

competitive access provider: (CAP, which see) Access services provided by a company other than a local exchange carrier (LEC), regional Bell operating company (RBOC), or AT&T, that is authorized to provide such service.

Competitive Telecommunications Association: (CompTel) An industry association of IECs that does not include AT&T, MCI or Sprint, but does include WorldCom and most medium-sized communications carriers. CompTel may also refer to one of the organization's conventions.

compression /decompression: A method of encoding/decoding signals that allows transmission (or storage) of more information than the media would otherwise be able to support. (for example, the "Stacker" software product more than doubles the storage capacity of a PC magnetic disk drive.) Both compression and decompression require processing capacity, but with many products, the time required is not noticeable. In faxes, one-dimensional compression is horizontal (Modified Huffman), two-dimensional compression is both horizontal and vertical, the latter compresses the space between the lines (Modified Read).

computer telephone integration: Multiple media hardware and software integrated for networking of information and workers as well as the enhancement of operation between the telephone and computer.

computer telephony: (term coined by Harry Newton) A philosophical idea, now enabled by electronics and hardware, that makes placing and receiving phone calls easier and more pleasing to corporate users at either end of the line. For additional information, see Computer TelephonyMagazine, the Miller Freeman website (www.computertelephony.com), or attend any of the Computer Telephony Conference and Expositions.

computer telephony integration: (CTI) The integration of telephony function with computer applications.

conference bridge: A telecommunications facility or service that permits callers from several diverse locations to be connected for a conference call.

conferencing capability: Allows several parties to speak from different sites on a single connection.

configuration: 1) The relative arrangement, options, or connection pattern of a system and its subcomponent parts/objects. 2) The process of defining an appropriate set of collaborating hardware and software objects to solve a particular problem.

connection: A point-to-point dedicated or switched communication path.

connection admission control: (CAC) The series of actions the network takes during the phase of call set-up to determine whether a connection can or should be made.

Construction and Maintenance Agreement: (CMA, C&MA) An agreement for the ownership, construction and maintenance of expensive facilities (such as transoceanic cables and related equipment). Such agreements are usually between multiple carriers, but may be between a carrier and a government.

contract: A legally-binding agreement between a vendor and a customer to provide products, services, or features in a specified quantity and quality, for a specified price, during a specified period of time.

contract carriage: The provision of regulated service pursuant to individually negotiated contracts, instead of through public tariffs.

contract tariffs: Services and rates based on contracts negotiated with individual customers, but theoretically available to all customers. AT&T has filed several hundred contract tariffs.

CONUS (or CON): The 48 contiguous United states. Used primarily to designate the operating range or authorization of a satellite or radio facility.

COPT: (Coin Operated Pay Telephone Correspondent) A local service provider in a country which exchanges traffic with a carrier. For example, British Telecommunication or Mercury could be the U.K. correspondent of a U.S. carrier.

COS: See Class Of Service.

country code: One, two or three digit codes used for international calls outside of the North American Numbering Plan area codes. From North America dial: 011 + country code + city code + local phone number. For example, in 011 + 91 + 22 + 123-4567, 91 = India, 22 =Bombay. Country and city codes are listed in the front matter of telephone directories in the US.

CPE: See Customer Premises Equipment

CPL: See Commercial Private Line

CR: (Customer Record) See also service management system (SMS).

CRC: (Cyclic Redundancy Check) A process that determines if the integrity of a block of data has been maintained during transmission, reading or writing of the data. A calculated number is appended to the transmission of the string of bits, and recalculated based on the received data. If the numbers match, there has been no error in transmission.

CREDFACS: (Conduit, Raceway, Equipment Ducts, and FACilitieS) Generic collective term

cross connect: A point in a network where a circuit is connected from one facility to another by cabling between the equipment.

CS: (Calling Seconds) A measure of communication traffic.

CSB: (Client Support Bulletin) NASC information to RespOrgs about NPA splits, etc.

CSID: (Calling [transmitting] Station Identification) The name/number field printed across the top of a fax sheet.

CSPDN: (Circuit Switched Public Data Network) Circuit oriented public network usually based on X.25.

CSU/DSU: (Customer Service Unit/Digital Service Unit) A device which is usually required for T-1 installations and leased from the long distance provider. It contains the last signal regenerator before the end-users equipment, allows circuit testing, monitors the data stream and provides power for the T-1. See Channel Service Unit /Data Service Unit

CTI: See Computer Telephony Integration

customer: An individual person or organization that purchases (orders, requests, or may be billed for) service. A customer may be related to an entity that pays for products. For example, a subsidiary company may have its own customer identification even though the parent company pays all charges. A billable customer may be someone that merely accepts an operator service call or a casual customer that dials a CIC code (like 10555) without presubscribing. A service provider or an agent may act as (or on behalf of) a customer. (Contrast with End User)

customer contact name: An SMS NUS NCON field. The designated person to notify as order status changes, etc. (Customer contact telephone number is in NUS NPHONE)

customer premises: The local facility where the circuit terminates.

customer premises equipment: (CPE) Communications equipment (such as PBX switches, origination/termination adapters, multiplexers, modems, codecs, telephones, computers, etc. - but not including carrier lines) at the customer's location that connects to carriers' products and services. CPE may be customer owned and maintained (COAM) or provided by the carrier. Primary CPE suppliers include AT&T, Northern Telcom, NEC, Phillips, Siemens, Erickson, and others.

customer record information system: (CRIS, pronounced "chris") A system used by many local exchange carriers (LECs) to maintain customer records.

customer type: Classification of customers that defines procedural rules and the availability of products, services, features and options (for example, residential, commercial, reseller, carrier, etc.)

cutover: The exact date/time that a phone number, circuit, etc. is scheduled to be (or was) moved from one implementation (carrier) to another. (For instance, moving an 800 number from MCI to WorldCom).

cut-through dialing: "10"+CIC+" #" followed by an AuthCode for IntraLATA calls.

D

D channel: (Delta channel) An ISDN interface. The out-of-band signaling link that carries packet-switched data using SS7 protocol. The D channel provides signaling information for the B channels.

DA: (Directory Assistance) The service that used to be called "Information." Usually accessed by pressing the area code and 555-1212, though there are variations. Charges for the service vary, too.

dB: (decibel) The logarithmic measure of signal strength or relative power between circuits, usually the difference between a transmitted signal and a standard signal source (from a standard signal generator). Decibels levels are expressed as the ratio of two values.

data link connection identifier: (DLCI)

debit card: A pre-paid long distance account. When long distance calls are made, the charges are deducted from this pre-paid account. When the account balance reaches zero, the telephone call is disconnected.

dialed number ID service: A feature that allows the person answering the phone to see the number that the caller dialed. This is particularly useful for businesses which handle details or services for several companies (catalog order-takers who work with catalogs from several businesses, for instance) and need to know which company was dialed.

dialup: A connection to a network, including the Internet, made by dialing or pushing numbers. Dialup can be accomplished by using a modem and a standard telephone to make a connection between computers.

DID Trunk/DID Extensions: (Direct Inward Dialing) The ability to call a internal extension without having to pass through an operator or attendant In large PBX systems, the dialed digits are passed down the line from the CO (central office). The PBX then completes the call. (See direct inward dial.)

digital loop carriers: (DLC) In general, a DLC is network transmission equipment that carries a voice-grade signal on multiple channels in a single 4-wire cable from the CO to a remote site. The cable from the CO to the remote site may be fiber-optic but at the customer end transmission generally goes into a twisted pair of copper wires.

digital signal: (DS) Digital signal speed is identified by numbers after DS, from DS-0 to DS-4, that indicate the capacity of lines and trunks. When used with a number, the DS is for Digital Service, that which is indicated by the various speeds and capacities. DS0, Digital Signal, level Zero, is 64 Kbps, the worldwide standard speed of digitizing a conversation using pulse code modulation. See DS-1.

digital subscriber line (or loop): (DSL) The technology enables fast digital communication of data and also provides an analog signal for voice. A cluster of digital services, collectively called xDSL, are provided to customers by their local phone company. The variety of services, some of which may not be widely available yet, include ADSL (Asymmetric DSL), CDSL (Consumer DSL), DSL Lite, HDSL (High-speed DSL), IDSL (ISDN [basic rate] DSL), MDSL (Medium-speed DSL), RADSL (Rate Adaptive DSL), SDSL (Symmetric DSL), VDSL (Very-high-speed DSL), and x2DSL.

digital technology: In telephone systems this term means that the inbound or outbound telephone calls are digitally transmitted from the telephone set to the telephone system. The advantages of digital telephone systems are 1) one pair of wires are needed to hook up a digital telephone, rather than up to four pair for a non-digital set; 2) the phone cannot be tapped between the phone system and the phone (a tapper would only hear a series of beeps).

direct inward dial: (DID) A local phone company provides this service that allows many phone numbers to ring on just a few lines. The phone company assigns a block of 100 phone numbers, 20 at a time. The company determines how many trunks or phone lines will be used to handle the calls. This service requires additional DID equipment that receives a signal from the phone company to indicate which phone number was dialed to make a phone line ring. The DID equipment passes the information to the switch or application which then can ring the correct extension.

DISA: (Direct Inward System Access) Allows an outside caller to dial directly into the telephone system and to access all the system's features and facilities. DISA is typically used for making long distance calls from home using a company's less expensive long distance lines, like WATS or tie lines.

discrete multi-tone: See DMT for description.

distinctive ringing: Different patterns or tones of ringing. Allows sorting of calls (by ear) among many users.

DNIS: (Dialed Number ID Service) which see.

do not disturb: (DND) Makes a telephone appear busy to any incoming calls. May be used on intercom-only, by extension line only, or both.

DLCI: (Data Link Connection Identifier, which see)

DMT: (Discrete Multi-Tone) Using digital signal processors, DMT divides the 1 MHz spectrum of a copper local loop into 256 4 KHz channels. The bit densities on each of the channels is automatically varied to minimize noise and interference. An advantage of DMT is that it can switch the data flow from a poor channel (one with interference) to a good one. ANSI and ETSI have established DMT as the standard for ADSL but CAP (Carrierless Amplitude and Phase) modulation, which is not standard, is often implemented side-by-side with DMT.

dpi: (dots per [square] inch) The measure of sharpness in scanning. More dots, sharper image.

DS: (Digital Signal) which see.

DS-0: (Digital Service, level zero) The basic digitizing speed of 64 Kbps.

DS-1: (Digital Service, level 1) 1.544 Mbps in North America, 2.048 Mbps everywhere else.

DS-1C: (Digital Service, level 1C) 3.152 Mbps in North America, carried on T-1.

DS-2: (Digital Service, level 2) 6.312 Mbps, carried on T-2.

DS-3: (Digital Service, level 3) 44.736 Mbps.

DS-4: (Digital Service, level 4) 274.176 Mbps.

DSL: (Digital Subscriber Line) which see.

DSL Lite: A slower version of ADSL in which the data and analog voice components are split by the CO for the user. Remote splitting reduces the maximum data rate to 1.544 Mbps (still higher than common modem data rates).

DTMF: (Dual Tone Multi Frequency) Technical term for the standard touch tone sounds generated by a telephone. Dual tone multi frequency, uses two tones to represent each key on the touch pad. When any key is pressed the tone of the column and the tone of the row are generated, hence dual tone. As an example, pressing the '5' button generates the tones 770Hz and 1336Hz. The frequencies were chosen to avoid harmonics (no frequency is a multiple of another, the difference between any two frequencies does not equal any of the frequencies, and the sum of any two frequencies does not equal any of the frequencies). The frequencies generated must be with +/- 1.5%, the highest frequency must be as loud as the lowest frequency and as much as 4db louder. This level difference is referred to as 'twist'.

E

E&M: (Ear and Mouth) E and M are signaling leads that accompany a voice path of 2 or 4 wires in a PBX. A PBX typically seizes a trunk by grounding the M lead (which appears as the E lead at the far end). Each PBX signals on its M lead and looks for signals on its E lead (remembered easily as Ear and Mouth, from the switch's point of view).

EAX: (Electronic Automatic Exchange) An electronic central office of a non-Bell phone company.

ECP: (Enhanced Call Processing) A voice mail system with the option of interactive voice response.

effective dialing date: When the new area code starts working.

EIR: (Equipment Identity Register or Excess Information Rate) see either.

EKTS: (Electronic Key Telephone System) A service available with ISDN PRI, which see.

electronic automatic exchange: (EAX) An electronic central office of a non-Bell phone company.

electronic switching system: (ESS) The term used by Bell and AT&T for an electronic central office system.

EMI: (Electromagnetic Interference) Radiation from electrical or magnetic fields that affect performance or responses of electronic equipment.

equipment identity register: (EIR) A database used to verify the ownership mobile phone equipment. Useful in case the equipment is stolen.

ESS: (Electronic Switching System) The term used by Bell and AT&T for an electronic central office system.

ethernet: A local area network used for connecting computers, printers, servers, etc. within the same building.

excess information rate: (EIR) EIR can usually be obtained from the network provider as a sort of bandwidth overdraft protection when both sending and receiving equipment is suitable. If arrangements are made that greater bandwidth is used during off-peak hours, the cost is considerably less.

external bus: A relatively simple electrical connection that allows calls between CT devices in the same system.

external paging access: Or rapid paging access can be made by an extension off the PBX by pushing one or several digits (or hitting a direct paging key). Attach external paging to the telephone system does not necessarily give every paging option. Sometimes a paging adapter is required to access several zones, or groups of speakers, or an "all call" (all speakers).

F

---

facsimile equipment, fax: Equipment which allows hardcopy (written, typed, or drawn material) to be sent through the switched telephone system and printed out elsewhere.

fast busy signal: Indicates that the call cannot be completed. Heard often with area code changes.

FaxBios: Standards developed by an industry consortium as an application programming interface (API) for fax boards.

fax broadcast: An automated method of faxing a single document to many fax numbers.

fax mail: Similar to voice mail except the system will store fax documents in individual mailboxes instead of storing voice files.

fax on demand: Fax technology coupled with voice technology to provide a means for callers to request information to be faxed back automatically without human intervention.

fax/modem card: Used for sending faxes to a fax machine or to another fax/modem card. The modem is used for data communications, for one computer to communicate with another computer and exchange data files.

FCC: (Federal Communications Commission) Independent federal agency, authorized by the Communications Act of 1934, responsible for services and common carrier activities that cross state lines (interstate traffic). Additionally, the FCC is responsible for regulating international telecommunications for the U.S. and is caretaker of all radio and TV broadcast regulation and radio frequency allocations.

FCS: (Frame Check Sequence) A part of the SDLC protocol that controls physical error by checking the integrity of a transmitted frame.

FDID: (Flexible Direct Inward Dialing)

FDM: (Frequency Division Multiplexing)

Feature Groups A, B, C, D: (FGA, FGB, FGC, FGD) Four different arrangements by which end-users can make toll calls.

FGA: (Feature Group A) Access to LEC is by subscriber-type line connection rather than a trunk.

FGB: (Feature Group B) Access to LEC is by trunk and call can be made from anywhere in the LATA.

FGC: (Feature Group C) Traditional arrangement with AT&T. Service includes automatic number identification, touchtone, and other AT&T amenities.

FGD: (Feature Group D) The equal-access service in which all interexchange (IX) carriers have the same connections to the local exchange. IX carrier is chosen by the customer and accessed by dialing or pressing "1" before the area code. With increased IX competition the other feature groups convert to FGD and the IX is billed by the LEC for measured use of the LEC's system.

flash hook: A signal sent by the voice mail system, required to transfer calls within a business office or to external phone numbers.

flexible intercept: Allows the assignment of operator intercept to identified extensions.

flexible restriction: Permits the phone system to disallow calls from certain extensions.

FOD: (Fax On Demand) See above.

FRAD: (Frame Relay Access Device) Needed to access a frame relay network.

frequency division multiplexing: (FDM) A technique of assigning each voice channel to a different 4 kHz frequency band. 24 voice channels total 96 kHz which is well within the capacity of a twisted pair. FDM technology has given way to digital signal processing which allows less noise and trouble with switching, controlling and maintaining physical channels.

FX: (foreign exchange) But foreign only to the central office. An FX is a tie line connected from one telco CO switch to that of another, presumably distant, Central Office, which is the Off Premises Exchange (OPX). When the line connects the CO switch to a PBX, the PBX is the OPX. The PBX may have its own OPX, in which case the X stands for extension-an extension that is part of the PBX system but not on the business premises.

G

gateways: Gateways connect two dissimilar communications networks. Gateways translate protocols, convert several media formats, and transfer certain information. See VoIP chapter.

Group III: The most popular international standard for fax transmission. Nearly every fax machine in the world supports the Group III standard. A newer standard, called Group IV, has been released but is not widely adopted.

graphics interface format: (GIF) A format that encodes images in bits so that the GIF file can be read by a computer and the picture be reproduced on the computer monitor.

H

handset: The part of a phone held in the hand to speak and listen. It contains a transmitter and receiver.

headset: A telephone transmitter and receiver assembly worn on the head.

HDSL: (High-speed Digital Subscriber Line) HDSL transmission is symmetrical; equal bandwidth is available for upstream and downstream transmission. Because the bandwidth is split equally between directions, the maximum data rate for HDSL is lower than that for ADSL.

HTML: (HyperText Markup Language) The language used to set up pages on the World Wide Web of the Internet. HTML is ASCII text enclosed by HTML commands in angle brackets. HTML documents have 3 elements: tags, define type styles and allow hyperlinks; comments, working words or symbols visible to the author but not on the finished product; text, the real content of the page.

hunting group: A hunting group, a service provided by the local phone company, consists of a group of phone lines which can all be reached by dialing a single main phone number. For example, each of 8 phone lines to a business has its own unique phone number. If all 8 of the lines are in a hunting group, then only the first number of the group need be advertised because if line one is busy, the phone company will hunt to the next available phone line in the group. Eight calls would have to be on the line simultaneously before callers get a busy signal.

hunting: Refers to the progress of a all reaching a group of lines. The call will try the first line of the group. If that line is busy, it will try the second line, then it will hunt to the third, etc. A Hunt Group refers to the telephone lines.

hybrid key system: Term used to describe a system which has attributes of both key telephone systems and PBXs.

I

IA: (Implementation Agreement) which see.

ICR: (Intelligent Character Recognition or Initial Cell Rate)

IDDD: (International Direct Distance Dialing) which see.

ILEC: (Incumbent Local Exchange Carrier)

implementation agreement: (IA) A working arrangement in the absence of agreed-upon international standards.

initial cell rate: (ICR) A service parameter of the ATM system that describes how many cells per second can be sent at the start of a transmission and after an idle period.

intelligent character recognition: (ICR) A system for "reading" fax and other printed characters into a word-processing program. It works better than OCR, but is far from perfect.

intermediate distribution frame: A metal rack designed to connect cables and located in an equipment room or closet. Consists of bits and pieces that provide the connection between inter-building cabling and the intra-building cabling, that is, between the main distribution frame (MDF) and the individual phone wiring.

internal bus: A system of connections that can have up to eight devices attached, each with its own internal phone number, and accessible from any of the other devices connected to the ISDN PBX.

international callback: Overseas customers are provided with a US dial tone through which calls can be originated at much lower rates. Once the service has been hired international callback works in 3 steps. 1) You call a number in the US, let it ring once and hang up. 2) A computer in the US calls you back at your offshore location, you enter a PIN number and the telephone call destination number. 3) A computer in the US originates a call from the US to your destination, then connects you through. Technically both calls (the one from the US to you and the one from the US to the destination number) originate in the US and are billed at much lower rates than international calls. Callback and interstate providers pay 3.9% of their international and interstate revenues to support the FCC's Universal Service Fund (USF).

international carriers: (INC) A carrier that provides connections from World Zone 1 (US, Canada, and some eastern off-shore islands and countries) to some of the rest of the 8 World Zones.

international direct distance dialing: (IDDD) Also called "international direct dialing" and "international subscriber dialing." An automatic switching system by which a caller can dial a specific number in another country. See discussion of "country codes" in the chapter on long distance services.

internet protocol: (IP) The most important of the protocols on which the internet is based. IP describes standard software that keeps track of internet addresses and routes of incoming and outgoing messages. The IP makes it possible for a packet of data to use many networks between origin and delivery.

IRQ: (Interrupt Request) A signal made in or by a PC memory card when a problem arises with hardware.

ISDN: (Integrated Services Digital Network) A totally new concept of what the world's telephone system should be. The network service comes at two levels and two speeds. BRI (basic rate interface) = 144 Kbps, PRI (primary rate interface) = 1544 Kbps.

ISDN BRI Service: (Basic Rate Interface) Brings two bearer channels and one D channel (2B+D) to your PC. Allows tremendous versatility with a good PC.

ISDN PRI Service: (Primary Rate Interface) Pretty much the "business" rate necessary to get all the computer telephony goodies. PRI service can set a business up with an electronic key telephone system (EKTS), an ISDN "centrex," and approximations of PBX service (and management) that is better, fancier and faster than any of them. And more expensive.

ISDN DSL: (IDSN [base rate] Digital Subscriber Line)

isochronous: The word means two-way transmission without delay. An ordinary conversation is isochronous.

ISP: (Internet Service Provider)

ITU-T: (International Telecommunications Union-Telecommunications) Located in Geneva, Switzerland, ITU-T is the primary international body for fostering cooperative standards for telecommunications equipment and systems.

IVR: (Interactive Voice Response) The system with a recorded list of options for the caller to respond to by pushing touchtone phone buttons.

IXC: (Interexchange Carrier) Long-haul long distance carriers. Inter-LATA carriers such as AT&T, Sprint, MCI, Worldcom, and a host of other, smaller companies.

J

jack: Receptacle used to connect a cord or line to a telephone system.

K

key telephone system/key system: (KTS) A system in which the telephones have multiple buttons permitting the user to select outgoing or incoming central office phone lines directly. Central office phone lines are used exclusively and the use of direct inward dialing (DID) trunks is not permitted. This is no longer strictly true; there are multiple hybrid systems available that can accommodate DID trunks. KTS systems are usual in small businesses that have about 50 phones and do not need the capabilities of a PBX. Electronic KTS provide switching and other functions and features.

KSU: (Key Service Unit) The "brains" of a key system.

LAN: (Local Area Network) A short distance network (typically within a building or campus) used to link together computers and peripheral devices (such as printers) under some form of standard control.

LATA: (Local Access and Transport Area) In nearly 200 such geographic areas in the US local phone companies may offer both local and long-distance service.

LCA: (Local Calling Area) which see.

LCD: (Liquid Crystal Display) An alphanumeric display using liquid crystal scaled between two pieces of glass. The display is divided into hundreds or thousands of individual dots, which are charged or not charged, reflecting or not reflecting external light to form characters, letters and numbers.

LCR: (Lease Cost Routing) A telephone system feature that automatically chooses the lowest cost phone line to the destination.

LEC: (Local Exchange Carrier) which see.

LD: (Long Distance or Loop Disconnect)

local calling area: (LCA) The area in which no tolls are charged for calls. Usually a toll is charged between LCAs. Metropolitan area phone companies may have plans by which an LCA is enlarged, but so is the monthly bill.

local exchange carrier: (LEC) The local phone company, either a Bell operating company (BOC) or a competitor. The former Bell companies are usually called ILEC (incumbent local exchange carriers) while the competing carriers are CLEC (competitive local exchange carriers). Strictly speaking, the former Bell companies are the ones meant when the word "telco" is used.

LNP: (Local Number Portability) In compliance with FCC criteria, a customer may switch phone companies and retain the same phone number.

LNPA: (Local Number Portability Administrator) Long-term number portability requires a national system of regional databases managed by an independent third-party local number portability administrator(s) (LNPAs) selected by the North American Numbering Council (NANC). Selection of LNPAs in certain areas has been slow, and legal holdups and technical unfeasibility are also contributing to LNP implementation delays.

logical link multiplexing: Allows voice and data frames to share the same permanent virtual circuit (PVC).

LRN: (Local Routing Number) The code of call routing within a local exchange company.

LSE: (local switching equipment) Switching equipment housed at the phone company's central office.

M

MAC: (Moves, Adds and Changes) Adjustments made to equipment on user premises.

main distribution frame: (MDF) The wiring structure which connects incoming wires from the phone company with the internal wires of a building.

MAPI: (Microsoft Application Programming Interface) which see.

MCI: Used to be an acronym for Microwave Communications, Inc. Now is just the company name. MCI International is the subsidiary that handles communications services to more than 100 countries outside the US borders.

MCR: (Minimum Cell Rate) which see.

MDF: (Main Distribution Frame) which see.

MDSL: (Medium-speed DSL) Medium-speed means less than the maximum (1.544 Mbps) of a T-1 digital transmission link.

message notification: A voice mail feature which calls a mailbox owner at a pre-determined number when a message is received in the mailbox and allows the box owner the option of listening to the messages at that time.

message waiting: A light on the phone or letters or characters on the phone's display indicating there's a message waiting somewhere for the owner of the phone.

MFJ: (Modified Final Judgment) The federal court ruling that defined the rules and regulations for the deregulation and divestiture of AT&T and the Bell system.

MH: (Modified Huffman) which see. A data compression method used in fax transmission.

Microsoft Application Programming Interface: (MAPI) Specifically this is the Windows interface for a messaging system that allows clients to interact with various other Microsoft and Windows NT message service providers. Using MAPI one can build a message handling system to fit personal needs.

minimum cell rate: (MCR) The minimum rate guaranteed a customer in an ATM system contract. A descriptor of an ATM service level.

MMR: (Modified Modified Read) which see.

MR: (Modified Read) which see.

modem: Acronym for modulator/demodulator. Equipment that converts digital signals to analog signals and vice-versa. Modems are used to send data signals (digital) over the telephone network, which usually is analog. A 28.8 Kbps modem is one that operates at 28.8 thousand bits (kilobits) per second. (Storage is measured in bytes; transmission capacity in bits per second.)

modified Huffman: (MH) A data compression method that squeezes the data horizontally only. Used in fax transmission of data. See "modified read."

modified modified read: (MMR) An improved data compression system used in newer Group 3 fax machines.

modified read: (read = relative element address differentiation) A 2-dimensional compression technique that squeezes fax data both vertically and horizontally, removing spaces in both directions.

modular: Equipment is said to be modular when it is made of "plug-in units" which can be added together to make the system larger, improve its capabilities, or expand its size. There are very few phone systems that are truly modular.

MSA: (Metropolitan Statistical Area) Based on US Census figures, MSAs are counties that contain cities of 50,000 population or more. The FCC uses MSA designations when it issues cellular licenses.

music on hold: (MOH) Background music heard while on hold. Lets callers know they are still connected. Please be careful about playing the radio or other non-licensed music . The FCC has ruled that it is illegal to re-broadcast, or play music without paying royalties to the artists. Specially-produced music including custom sales tapes by another company that has paid royalties is OK.

muxes: More than one multiplexer.

N

NAA: (Next Available Agent) A feature of large phone systems, well-suited to customer service or catalog sales setups, in which an incoming call is routed to the person/agent whose phone line is not in use.

NANC: (North American Numbering Council) The folks who administer the NANP and select local number portability administrators (LNPAs) to implement the FCC rules on number portability. The council members are drawn from the carrier, manufacture, and end-user communities.

NANP: (North American Numbering Plan) which see.

network interface card: (NIC) The enabling connection between a computer or other device and a cable or other medium shared by other stations.

network node interface: (NNI) An ATM term for the interface between pieces of equipment belonging to different public frame relay networks.

network termination: (NT) When incorporated into user equipment, NT converts analog signal to ISDN and vice versa. NT1 and NT2 are ISDN adapters that support analog telephone devices such as phones, data, modems, Group 3 faxes, and answering machines.

network-to-network interface: (NNI) A protocol defined by both the Frame Relay forum and the ATM forum for routing of signals. Includes network node interface (NNI) and user-to-network interface (UNI) protocols.

NIC: (Network Interface Card) which see.

night answer: Incoming calls to a switchboard during evening and weekend hours are automatically rerouted to ring only at designated night answering phones such as the security desk.

NNI: (Network Node Interface or Network to Network Interface) which see.

North American numbering plan: (NANP) The method of identifying telephone trunks in the public network. Identification is made by area code, exchange code, and the 4-digit subscriber number. (The US and Canada together are World Numbering Zone 1.)

NRM: (Normal Response Mode) A master-slave mode in which only one station at a time can transmit frames.

NT: (network termination) which see.

numbering plan area: (NPA) The long way to say "area code."

O

ODBC: (Open Database Connectivity) A standard used by many software programs including Excel and Microsoft Word. ODBC allows import/export data in/out of other formats so the information can be viewed in a preferred form, without the user having to learn new software.

OCR: (Optical Character Recognition or Outgoing Call Restriction)

octothorpe: The pound sign on a telephone keypad.

OEM: (Original Equipment Manufacturer)

open numbering plan: The numerical system used for international dialing. It can handle 7 to 15 digits that include one of the 8 world numbering zones (Zone 1 includes the US, Canada, and numerous Caribbean countries and islands), a one-to-three-digit country code, within-country area codes, local exchange numbers, and finally the subscriber's own identifying number.

optical character recognition: (OCR) How a machine scans the characters in a document and converts the data into a standard form than can be manipulated and stored electronically. In other words, a conversion of data (letters, numbers) on faxes and other printed material to a form suitable for computer processing.

OPX: (Off Premises Extension) Most frequently a PBX extension not on the business premises.

OPX: (Off Premises Exchange) The far terminus of a trunk line between two telco central offices.

outbound call queuing: Arranges outgoing calls in line for the first available trunk.

outgoing call restriction: (OCR) An option that allows only in-coming calls.

P

---

packets: A packet, as used in data transfer systems, consists of a header, signaling bits and a payload (an information field).

pager notification: A voice mail feature that calls a mailbox owner's pager number to announce a message in the mailbox. Most voice mail systems can ask the caller to enter a phone number and will then display that number on the mailbox owner's pager.

paging: To give a message to someone who is somewhere else. Paging can be done with a little "beeper" carried in on one's person. Paging can also be done through speakers in phones, or from speakers in the ceiling. Most phone systems offer a paging channel access. A number can be dialed and the person paged.

PBX: (Private Branch Exchange) A private phone switching system, usually on a customer's premises with an attendant console, that allows communication within a business and between the business and the outside world. It is connected to a common group of lines from one or more central offices to provide service to a number of individual phones. A PBX differs from a key system in that to make an outgoing call on a PBX, the number 9 must be dialed first.

PBX trunk: A circuit which connects the PBX to the local telephone company's central office switching center or other switching system center.

PCA: (Protective Coupling [or Connecting or Connective] Arrangement) A device that AT&T/Bell historically required to be installed, as a rental unit, between AT&T/Bell lines and non-AT&T/Bell equipment. The FCC later ruled that the device was unnecessary and AT&T/Bell refunded most rental payments.

PCM: (Pulse Code Modulation) Encoding an analog voice signal into a digital bit stream.

PCR: (Peak Cell Rate) which see.

peak cell rate: (PCR) The rate of sending data that must not be exceeded by an ATM customer. Peak rates are not usually sustainable but may be used for relatively long periods when traffic is light.

permanent virtual circuit: (PVC) As part of a frame relay network, a PVC behaves like a private line but in fact uses preprogrammed routes from one point to another on demand.

permissive dialing end date: The last day the existing area code works. After that only the new area code works.

permissive dialing period: Time between the effective dialing date and the permissive dialing end date, the period in which you can use either the old or new area code.

PIC: (Primary Interexchange Carrier or Presubscribed interexchange carrier) which see.

PIM: (Personal Information Manager) A PC software program that allows great control, sorting, and archiving of fax, email, voicemail, etc. (See Unified Messaging chapter.)

PIN: (Personal Identification Number)

pixel: (picture element) A single small point on a cathode ray tube (CRT) display or a fax transmission. Pixels can be turned off or on and individually colored.

Plexar: See Centrex.

point-to-point protocol: An 8-bit serial interconnection protocol often used between a computer and the Internet. PPP can also connect two computers over ISDN.

polling: Periodic access to a device for data retrieval (or other purposes). Call accounting systems are frequently set up to "poll" a CDR/SMDR capture buffer attached to a PBX at regular intervals; retrieving the information it contains and subjecting it to analysis.

power failure transfer: In the absence of back up power from batteries or emergency generators, this feature switches to single line phones that draw their power from the phone lines themselves.

PPP: (Point-to-Point Protocol) which see.

presubscribed interexchange carrier: (PIC) An option offered by the local telco by which a subscriber can select a long distance carrier. Long distance carriers pay an access charge to the local telco. If a subscriber doesn't choose (presubscribe) a LD carrier, one will be assigned by the local telco and the subscriber and the LD carrier may be billed for the PIC charges.

primary interexchange carrier: (PIC) The long distance carrier to which a call is automatically routed (after the caller has dialed "1" before the number) in equal access areas. The local telco assigns a PIC code number to all local subscribers. When a subscriber changes long distance carriers, the new PIC code is assigned to the subscriber's phone.

privacy: Privacy usually means that once a caller "seizes" a line, no other user can access that same line even though it appears on his/her key set.

private line: A direct channel specifically dedicated to a customer's use between specified points.

protective coupling (or connecting or connective) arrangement: (PCA) A device that AT&T/Bell historically required to be installed, as a rental unit, between

AT&T/Bell lines and non-AT&T/Bell equipment. The FCC later ruled that the device was unnecessary and AT&T/Bell refunded most rental payments.

PSTN: (public switched telephone number) The local phone company.

PTT: (Post Telephone and Telegraph) Telephone systems usually administered and controlled by governments of non-US countries.

PUC: A state governmental agency that defines various policy making and oversees functions regarding regulated activities, such as telecommunications, within their state. The term PUC can vary by state (e.g., a state may have this agency entitled the Board of Public Utilities (BPU) etc...).

pulse or rotary dial: Pulse telephones send a series of clicks instead of dual tone multi frequency (DTMF, a combination of high and low frequency tones) touch tones to communicate with the voice mail system. Pulse-to-tone converters are needed to support both pulse and rotary telephones.

PVC: (Permanent Virtual Circuits) which see.

Q

---

quadrature amplitude modification: (QAM) A modulation technique that uses variations in signal amplitude; used in V.34 (28.8 Kbps) modems.

QoS: (Quality of Service) Although QoS is hard to measure in analog voice transmission (static? loudess of signal? clarity?) it is much better defined, and by experts, for the emerging ATM system in which digital packets are transmitted. QoS of ATM service includes cell error ratio, severely eroded cell block ratio, cell loss ratio, cell misinsertion rate, cell transfer delay, mean cell transfer delay and cell delay variability (each of which has an acronym). Based on the foregoing parameters, four Class of Service (CoS) designations have been defined (see CoS).

R

RA: (Return Authorization) A code number issued to a buyer to facilitate the return of a product for repair, replacement or refund. The number must be visible on the outside of the package being returned.

RADSL: (Rate-Adaptive DSL) A rate-adaptive ADSL technology (Westell) in which the equipment determines the rate signals can be transmitted over customer loops and adjusts the delivery rate accordingly.

rasterizing: A process by which a document image is converted to a stream of bits representing either black, white, or one of 16 levels of gray for each element of the image.

redundancy: 1) That part of any message which can be eliminated without losing the important information. 2) Having one or more "backup" systems available in case of failure of the main system.

remote call forwarding: (RCF) A service that allows a customer to have a local telephone number in a distant city. Every time someone calls that number, that call is forwarded to the distant customer.

remote programming: Allows changing the telephone system's programming remotely by dialing a phone system from a personal computer, modem and a communications software package.

RFI: (Radio Frequency Interference) Poor performance of electronic equipment caused by radio waves.

RFI: (Request For Information) An announcement of intent to purchase equipment that is sent to vendors, requesting general information about the equipment/systems available. Does not include a request for prices, which is the second step in the process and is done in an RFQ (which see).

RFP: (Request for Proposal) A detailed document prepared by a buyer defining his requirements for service and equipment sent to one or several vendors. A vendor's response to an RFP will typically be binding on the vendor. The vendor will be obliged to deliver the goods and services stated in the RFP at the prices, and following the conditions explained in that RFP.

RFQ: (Request For Quotation) The second step in selecting vendor of equipment. An RFQ, sent to several vendors who have responded to an RFI, contains a general description of the equipment/system the buyer intends to purchase (presumably after the buyer has done some research) and asks for an approximate quotation. The vendor may have good suggestions to offer, so a wise buyer may list only the configuration of a desirable system and the features that are important for his/her business.

RMA: (Return Materials Authorization) Same as RA (see above).

roaming: Subscriber is "roaming" in another cellular network. Roaming ANI identifies the mobile directory number (DN) placing the call, but does not necessarily identify the true forwarded-call origin.

S

SAC: (Service Access Code) which see.

SCR: (Sustainable Cell Rate) which see.

SDLC: (Synchronous Data Link Control) A data communications line protocol associated with IBM systems network architecture (which see).

SDSL: (Symmetric DSL) Upstream and downstream bandwidths are the same, with a range of 10,000 feet. In the US the rate is 1.544 Mbps, in Europe it is 2.048 Mbps.

server: A computer providing a service, such as shared access to a system, a printer, or an electronic mail system to LAN users.

service access code: (SAC) A SAC is a 3-digit code, similar to an area code and used as the first 3 digits of a 10-digit telephone number. The numbers 600, 700, 800 and 900 have been assigned to identify generic services or to allow access.

service profile identifier: (SPID) An additional number given to each telephone on a new ISDN line. The 8- to 14-digit SPID encodes the services ordered. SPID numbers are in the phone company database and are checked each time the ISDN line is used. Harry Newton suggests that an ISDN line "owner" tattoo the SPID numbers on some portion of his/her anatomy so as to have them unfailingly available (and presumably visible) when dealing with the phone co, which is reluctant to divulge one's SPID.

SIG: (SMDS Interest Group) Now defunct because SMDS has been pushed out of the market by Frame Relay and ATM technology. See switched multimegabit data service.

signaling system 7: (SS7) which see.

SLIP: (Serial Line Internet Protocol) The protocol used to run the Internet protocol over phone circuits.

smart jack: A smart jack is an interface in the wiring of a main distribution frame (MDF). The jack is not part of the customer premises equipment. It is owned and supplied by the phone company.

SMDI: (Station Message Detail [or Desk] Interface) A data line from the central office that gives information and instructions to your Centrex voice mailbox.

SMDR: (Station Message Detail Record). Synonymous with CDR.

SMDS: (Switched Multimegabit Data Service) which see.

SNA: Systems Network Architecture) which see.

SOHO: (Small Office, Home Office) A growing market for phone systems as more people work from their homes.

speed dial: A feature that enables a PBX or PBX phone to store certain telephone numbers and dial them automatically when a code is entered.

speed dialing: Permits fast dialing of frequently used numbers. A repertory of numbers may be stored in the instrument (telephone) and/or in the telephone switch,. Usually a button or one, two or three digits are dialed to activate speed dialing.

SPID: (Service Profile Identifier) which see.

SPLD: (Service Provider ID) Caller ID provided by the telephone company.

SS7: (Signaling System 7) A many-layered protocol that must be included in central office switch software if IP telephony is to reach its full potential.

standard business lines: Basic dial tone service supplied by the central office at business, as opposed to residential, rates.

station cable: The cable that is part of the horizontal cabling subsystem. In other words, the cable from either the telephone system to the telephone, or from the IDF (Intermediate Distribution Frame) to the telephone.

subchannel multiplexing: A technique used to combine several voice conversations within the same frame. Multiple voice payloads in the same frame allow increased efficiency and lower overhead.

sustainable cell rate: (SCR) A descriptor of ATM traffic management that indicates the rate/speed of transfer that can be maintained over a period of time. The sustainable cell rate is lower than the peak cell rate (PCR) and higher than the minimum cell rate (MCR).

switched multimegabit data service: (SMDS) A public network service intended to provide high-speed (45 Mbps) data transmission, generally LAN-to-LAN. Although the service had many advantages and was designed to fit with ATM, it failed to gain broad popularity.

switched virtual circuits: (SVC) Provide connections for the duration of the data transfer over the best path in the network (as determined by traffic) when the call is placed.

systems network architecture: (SNA) An early IBM system, widely used in mainframe computers. It consists of specifications, a plan for the network and a set of products. SNA establishes a logical path between network nodes, routing each message with address information contained in the protocol. The protocol can be updated and modified to work with more advanced systems. SNA is the earliest "layered" model but has also been called "tree-structured." The backbone of the SNA system consists of the system services control points (SSCP), a front-end processor (FEP) channels and SDLC lines.

T

T-1 (or T1): (Transmission-1) A digital transmission link with a capacity of 1.544 Mbps (1,544,000 bits per second). T-1 uses two pairs of normal twisted wires, like household wiring. T-1 normally can handle 24 voice conversations, each one digitized at 64 Kbps.

TAPI: (Telephone Application Programming Interface) Refers to the Windows telephony application programming interface (API). TAPI is a changing (improving) set of functions supported by Windows that allows Windows applications (Windows 3.xx, 95 and NT) to program telephone-line-based-devices (both digital and analog), modems, and fax machines in a device-independent manner. TAPI simplifies the process of writing a telephony application that works with a wide variety of modems and other devices supported by TAPI drivers.

TCPA: (Telephone Consumer Protection Act) The 1991 Act restricts certain types of unsolicited telephone calls

TCP/IP: (Transmission Control Protocol/Internet Protocol) A networking protocol that provides communication across interconnected networks, between computers with diverse hardware architecture, and various operating systems (Microsoft definition). The family of common internet protocols.

TDM: See Time Division Multiplex

TE: (Terminal Equipment) Any user-owned hardware (phone, fax, etc.) capable of working with ISDN.

TE-1: (Terminal Equipment, Type 1) Any ISDN standard equipment that can operate when connected directly to the ISDN network

TE-2: (Terminal Equipment, Type 2) Equipment that is not up to ISDN standards (an analog phone, for instance) and therefore needs a terminal adapter at the interface.

Telecommuting: The process of commuting to the office through a communications link rather than transferring one's physical presence.

telemanagement system: A modular, multifunction database-management system that usually incorporates features for PBX maintenance, call accounting, toll fraud detection, directory services, and so forth.

Telephone Consumer Protection Act: (TCPA) A 1991 federal law that restricts certain types of unsolicited telephone calls.

terminal adapter: (TA) Terminal adapters connect to ISDN an circuit to provide 128 Kbps internet access or data transmission. A terminal adapter sends digital codes (zeroes and ones). A TA is not a modem.

terminal equipment: (TE) Any user-owned hardware (phone, fax, etc) on the user side of an ISDN system.

test number: The number you dial to test whether your phone knows the new area code. Each new area code has its own test number.

text-to-speech: (TTS) Conversion of text to synthetic speech.

tie line: A dedicated circuit linking two points without having to dial the normal phone number.

time division multiplex: (TDM) A system of interspersing short pieces of several different data transfers into a single stream, and doing it so quickly that none of the recipients note the gaps in individual "messages."

toll fraud/telabuse: Subversion of a phone system for purposes of stealing long-distance or other phone service. "Inappropriate" use of a phone system (e.g., for calls to Astrology Hotline) by employees.

toll free number: A telephone number in the 800 or 888 area codes. When one of these numbers is dialed, the caller pays no long distance charges. Instead, the cost of the call is billed to the owner of the toll free number. International toll free numbers are also available. These international numbers often have different area and country codes.

toll restriction: Curbs a telephone user's ability to make long distance calls.

tone sender units: (TSU) A device inside the telephone system that sends the touch tone signals to circuits on an interface or station card inside of a PBX system.

touch tone: See DTMF.

trunk: A communication line between two switching system. Typically the trunk includes equipment in a central office and PBXs.

TSAPI: (Telephony Server Application Programming Interface) Described by AT&T, its inventor, as "standards-based (applications programming interface) API for call control, call/device monitoring and query, call routing, device/system maintenance capabilities, and basic directory services."

TSU: (Tone Sender Units) which see.

TTS: (Text To Speech) Conversion of text to synthetic speech

TTY: (TeleTYpewriter) Sends alphanumeric information over telecom networks.

TTY/TTD: (text telephones) typewriter-like communication devices that permit individuals with speech or hearing disabilities to communicate by typing messages back and forth over telephone lines. Arrangements can be made for a TTY/TTD user to communicate with regular telephones and hearing, speaking persons.

turnkey system: An entire telephone system with hardware and software assembled and installed by a vendor and sold as a total package.

twisted pair: Two insulated copper wires twisted around each other to reduce induction (thus interference) from one wire to the other. The twists, or lays, are varied in length to reduce the potential for signal interference between pairs. Several sets of twisted pair wires may be enclosed in a single cable. In cables greater than 25 pairs, the twisted pairs are grouped and bound together in a common cable sheath. Twisted pair cable is the most common type of transmission media.

U

UNE: (Unbundled Network Element) Services, training, or software sold separate from hardware systems.

UNI: (User-to-Network Interface) which see.

uninterruptible power supply: (UPS, pronounced "ups") An auxiliary power unit for a telephone system that provides continuous power in case commercial power is lost.

Universal Service Fund: (USF) Established by the FCC, administered by the National Exchange Carrier Association, the USF is a mechanism for cost allocation that keeps local exchange rates at reasonable levels, especially in rural areas.

UPS: (Uninterruptible Power Supply) A system of batteries or a generator that provides electrical power when the common power transmission fails.

user-to-network interface: (UNI) The actual contact point between a user's equipment, a FRAD, and the public network service provider, a frame relay or ATM service.

V

VAC: (Volts Alternating Current) A measure of an electrical current that reverses direction at regular intervals. Normal US electrical supply reverses direction 60 times per second (60-cycle current).

value added reseller: Vendors that package products with software for specific industries.

VAR: (Value Added Reseller) which see.

variable bit rate: (VBR) An ATM voice service in which voice conversations are limited to the bandwidth they need while the remaining bandwidth is given over to other services that may need it at that moment. VBR is also used when messages are generated in a bursty, random manner rather than steadily.

VBR: (Variable Bit Rate) which see.

VDC: (Volts Direct Current) A measure of electrical current that moves always in one direction. Batteries produce direct current.

VDSL: (Very-high-speed DSL) VDSL has an upstream bandwidth of 1.6 to 2.3 Mbps upstream bandwidth, a downstream bandwidth of 13 to 52 Mbps, and a 1,000 to 4,500 foot range. Over a short distance, as in a local area network, transmission by VDSL is excellent. Longer distances may require an optical fiber. The technology and standards for it are under development.

virtual private networks: (VPN) A connection given priority by a long distance carrier so that it is almost like a dedicated, point-to-point line. VPN have pre-programmed routes through the network from source to destination.

voice activity compression: (VAC) Conserving transmission capacity by eliminating pauses in speech.

voice grade: A communications channel which can transmit and receive voice conversation in the range of 300 Hertz to 3000 Hertz.

voice grade: A communications channel operating in the range of 300 to 3000 Hz.

voice mail card: Used as an interface for humans to be able to communicate with a computer by using a standard touch tone phone.

voice mail system: A device to record, store and retrieve voice messages. One type of voice mail device "stands alone" and the other is integrated with the user's phone system. A stand alone voice mail is not dissimilar to a collection of single person answering machines, with several added features. The machines (voice mailboxes) can be instructed to forward messages among themselves.

Voice mail can be organized to allocate many separate mail boxes so individuals can dial, leave messages, pick up messages from, pass messages along, and so forth. A user can also edit messages, add comments and deliver messages to a mailbox at a prearranged time. Messages can be tagged "urgent" or "non-urgent" or stored for future listening. The range of voice mail options varies among manufacturers. An integrated voice mail system includes two additional features. First, it signals the presence of messages by illuminating a light on the phone set and/or puts a message on the phone's alphanumeric display. Second, if the phone rings a certain number of pre-set times, the caller will automatically be transferred to the voice mail box, which will answer the phone, deliver a little "I am away" message and then receive and record the caller's message

voice over: A feature on a business phone system that allows the operator to talk to an employee making a call and be unheard by the person at the other end of the employee's call.

VoFR: (Voice over Frame Relay) An arrangement of virtual circuits that can carry both voice (much compressed) and data at the same time.

VoIP: (Voice over the Internet Protocol) A computer telephony technology. Voice is transmitted over a packet-based data network using Internet Protocol (IP). Not necessarily the same thing as Internet Telephony which uses the public Internet.

volt: The unit of electrical potential and electromotive force equal to the difference in potential needed to cause a current of one ampere to flow through the resistance of one ohm.

voltamp: (VA) volts times amps.

VPN: (Virtual Private Network) which see.

W

WATS: (Wide Area Telecommunications Service) WATS is a now-generic name for discounted long distance toll service. Incoming WATS calls are usually to 800 or 888 numbers.

watt: The unit of electricity consumption. (A unit of power equal to one joule per second.)

World Zones: The eight numbered geographical divisions that permit assigning each subscriber an individual telephone number. (Zone 1 includes the US, Canada, and several nations and island groups off the southeastern shore of North America.)

X

xDSL: (x = generic Digital Subscriber Line) The technologies involved in DSL are collectively called xDSL, and enable fast (high bandwidth) digital communication of data and an analog signal for voice communication.

GLOSSARY

xDSL networks: The networks have 5 parts: digital subscriber line access multiplexers (DSLAM) at the network end, and xDSL modems, plain old telephone system splitters (POTS), channel service unit/digital service unit (CSU/DSU), and Ethernet network interface cards (NIC) in connected PCs at the customer end.

x2/DSL: This is an upgradeable modem that the manufacturers hope will replace xDSL modems and go far beyond. The x2/DSL modem now supports 56 Kbps communications.